Mathematical and Analytical Techniques
with Applications to Engineering

The importance of mathematics in the study of problems arising from the real world, and the increasing success with which it has been used to model situations ranging from the purely deterministic to the stochastic, in all areas of today's Physical Sciences and Engineering, is well established. The progress in applicable mathematics has been brought about by the extension and development of many important analytical approaches and techniques, in areas both old and new, frequently aided by the use of computers without which the solution of realistic problems in modern Physical Sciences and Engineering would otherwise have been impossible. The purpose of the series is to make available authoritative, up to date, and self-contained accounts of some of the most important and useful of these analytical approaches and techniques. Each volume in the series will provide a detailed introduction to a specific subject area of current importance, and then will go beyond this by reviewing recent contributions, thereby serving as a valuable reference source.

More information about this series at http://www.springer.com/series/7311

Rafael Martínez-Guerra ·
Oscar Martínez-Fuentes ·
Juan Javier Montesinos-García

Algebraic and Differential Methods for Nonlinear Control Theory

Elements of Commutative Algebra and Algebraic Geometry

Springer

Rafael Martínez-Guerra
Departamento de Control Automático
CINVESTAV-IPN
Mexico City, Distrito Federal, Mexico

Oscar Martínez-Fuentes
Departamento de Control Automático
CINVESTAV-IPN
Mexico City, Distrito Federal, Mexico

Juan Javier Montesinos-García
Departamento de Control Automático
CINVESTAV-IPN
Mexico City, Distrito Federal, Mexico

ISSN 1559-7458   ISSN 1559-7466 (electronic)
Mathematical and Analytical Techniques with Applications to Engineering
ISBN 978-3-030-12027-6   ISBN 978-3-030-12025-2 (eBook)
https://doi.org/10.1007/978-3-030-12025-2

Library of Congress Control Number: 2018968389

This Springer imprint is published by the registered company Springer Nature Switzerland AG
The registered company address is: Gewerbestrasse 11, 6330 Cham, Switzerland

Preface

This book is written so the reader with little knowledge on mathematics and without experience in control theory can start from the beginning and go directly through the logical evolution of the topics. If the reader has advanced knowledge on mathematics, is fluent on abstract algebra, linear algebra, etc., will be able to read this book a little faster, even so, the reader must study the mathematics sections of the book, to make sure he is familiar with any special notation that we have presented. If the reader only knows about mathematics but has no experience in natural sciences or engineering, he must not be discouraged for the occasional use of examples and vocabulary of other fields (physical models or others). If the reader knows a little about control theory, maybe from previous courses of engineering or physics, it is possible that he will have to work slightly more at the beginning to get used to the ideas and notations in modern mathematics. When the reader is in a section that contains some mathematics that are new to him, it is required of the reader to study the examples, to do many exercises and to learn the new subject as if he has taken a course of it. At the end of the book, the reader will know not only control theory but also a little about useful mathematics. Whatever the previous experience the reader has, he will find in the book some topics that are needed to be learned for this particular area. As well, the book shows in a simple way some elementary concepts of mathematics such as basic algebraic structures (sets, groups, rings, fields, differential rings, differential fields, differential field extensions, etc., that is to say elements of commutative algebra and algebraic geometry in general) for the introduction to the theory of nonlinear control seen from a differential and algebraic point of view. Enough material has been included, rigorous proofs and illustrative exercises of each subject, as well as some examples and demonstrations are left to the reader for the good understanding of the chapters of the book. In the selection of material for this book, the authors intended to expound basic ideas and methods applicable to the study of differential equations in theory of control. The exposition is developed in such a way that the reader can omit passage that turn out to be difficult for him. The content of this book constitutes a tool that can be read by not only mathematicians, engineers and physicists but also all users of the theory of differential equations and theory of control. Control theory has become into a scientific discipline and of engineering that seems destined to have an impact in all aspects of modern society. First studied by mathematicians and engineers now it is increasingly present on areas of physics, economy, biology, sociology, psychology, etc. What is control theory and why is so important for a wide variety of special-ities? Control theory is a set of concepts and techniques that are used to analyze and design various kinds of systems, independently from their physical nature and special functions. The most important aspect of control theory is the development of a quantitative model that describes the relation and interaction of a cause and effect between the variables of a system. This means that the language of control theory is mathematical and any serious attempt to learn control theory must be accompanied of mathematical precision and its comprehension. We have attempted, at writing this book, to make a pedagogic introduction to control theory that is accessible to readers with little background beyond linear and abstract elemental algebra and matrix theory while begin motivating for those mathematician readers with more sophisticated theoretical knowledge. As result, the book is more autonomous and less specialized. This book begins with the study of elementary set and map theory being a little bit abstract. Chapters 2 and 3 on groups theory and rings respectively are included because of their important relation to the linear algebra, the group of invertible linear maps (or matrices) and the ring of linear maps of a vector space being perhaps the most amazing examples of groups and rings. We deal with homomorphisms and Ideals as well as in Chap. 3 we mention some properties of integers. Also it is well known that group theory at its inception was deeply con-nected to algebraic equations. The success of the theory of groups in algebraic equations led to the hope that similar group-theoretic methods could be a powerful arsenal to attack the problems of differential equations. Still more, Ritt and Kolchin tackled the problem form the differential and algebraic point of view (Chap. 10 of control theory from point view of differential algebra). Chapter 4 is devoted to the matrices theory and the linear equations systems. Chapter 5 gives some definitions of permutations, determinants and inverse of a matrix. In Chap. 6 is tackled in notion of vector in real Euclidian space and in general vector space over a field, bases and dimension of a vector space as well as sums and direct sums. In Chap. 7 we treat with linear maps or linear transformations, Kernel and image of a linear map, dimensions of kernel and image, in addition, the application in linear control theory of some abstract theorems such as the concept of kernel, image and dimension of space are illustrated. Chapter 8 we shall consider the diagonalization of a matrix and the canonical form of Jordan. In Chap. 9 we attack elementary methods of solving differential equations and finally in Chap. 10 we treat the nonlinear control theory from the point of view of differential algebra.

Mexico City, Mexico   Rafael Martínez-Guerra
                                                                                Oscar Martínez-Fuentes
                                                                         Juan Javier Montesinos-García

Acknowledgements

To the memory of my father
who was the source of my teaching
Carlos Martínez Rosales.

To my wife and Sons
Marilen, Rafael and Juan Carlos.

To my mother and brothers
Virginia, Victor, Arturo, Carlos, Javier
and Marisela.

Je tiens à exprimer ma profonde gratitude
au Dr. Michel Fliess qui m'a initié au domaine
du contrôle non linéaire à l'aide, d'outils
tels que l'algèbre différentielle. Merci
également à mon ami Dr. Sette Diop
avec qui j'ai eu le plaisir de collaborer à certaines
publications. à eux ma plus sincère gratitude.

Rafael Martínez-Guerra

To my parents Teresa and Leopoldo.
In memory of my grandfather, Facundo.

Oscar Martínez-Fuentes

To my parents Irma and Javier,
whose support and love allowed me
to reach many goals.

To my aunt Rosario and uncle Clemente,
for all their teachings and support.

Juan Javier Montesinos-García

Contents

Notations and Abbreviations

$\mathbb{N}$	The Set of natural numbers
$\mathbb{Z}$	The Set of integers numbers
$\mathbb{Q}$	The Set of rational numbers
$\mathbb{R}$	The Set of real numbers
$\mathbb{C}$	The Set of complex numbers
$A, B, \ldots$	Capital letters represent arbitrary sets
$x, y, \ldots$	Lowercase letters represent elements of a set
$A \subset B$	A is subset of B
$x \in A$	x is element of A
$A \cup B$	The union of two sets
$A \cap B$	The intersection of two sets
A^c	Complement set of A
$A \backslash B$	Difference of sets A and B
$\emptyset$	Empty set
$\Longleftrightarrow$	Necessary and sufficient condition
$\forall$	For all
$\sim$	Equivalence relation
$a \equiv b \bmod n$	a is congruent with b module n
A/R	Quotient set
$*$	Binary operation for groups
$\det(A)$	The determinant of a square matrix $A \in \mathbb{R}^{n \times n}$
$|A|$	The determinant of a square matrix $A \in \mathbb{R}^{n \times n}$
$\mathrm{tr}(A)$	The trace of a matrix A
$A^{\top}$	The transpose of a matrix A
$\{\cdots\}$	Set
$(a_{ij})_{i,j}$	$m \times n$ matrix with entries $a_{ij}, 1 \leq i \leq m, 1 \leq j \leq n$
$\mathrm{rank}(A)$	Rank of a matrix A
A^{-1}	Inverse of A
$\dot{y} = \frac{dy}{dt}$	First derivative of y with respect to t
E_λ	Eigenspace of E corresponding to eigenvalue λ

$\dim(V)$	Dimension of V
$\mathscr{L}(V, W)$	Space of linear transformations of V in W
Ker φ	Kernel of φ
diff trd°	Differential transcendence degree
$\square$	Designation of the end of a proof
$<(\,>)$	Less (greater) than
$\leq(\geq)$	Less (greater) than or equal to
$\forall$	For all
$P(n)$	A statement
$\cong$	Isomorphism
G/N	Quotient group
$K\langle u \rangle$	Differential field generated by the field K, u and its time derivatives
K/k	Differential field extension $k \subset K$

Chapter 1
Mathematical Background

Abstract This chapter focuses on the basic concepts and algebra of sets as well as a brief introduction to the theory of functions and the well known principle of mathematical induction. All this background will be needed as a tool to understand the theory of linear algebra and differential equations as set forth in the following chapters.

1.1 Introduction to Set Theory

Definition 1.1 A **set** A is any collection of objects called **elements** of A that is well defined (same nature) [1–4]:

$$A = \{a_1, a_2, a_3, \ldots, a_{n-1}, a_n\}$$

- $a_i \in A$ denotes a_i is an element of set A.
- $x \notin A$ denotes x is not an element of set A.

Hereinafter, capital letters, $S, T, \ldots$ denote sets and each of the elements of a set is denoted by a lowercase letters, $x, y, \ldots$, etcetera.

Example 1.1 Let $\mathbb{N} = \{1, 2, \ldots\}$ the set of natural numbers (positive integers), $\mathbb{Z} = \{\ldots, -3, -2, -1, 0, 1, 2, \ldots\}$ the set of integers, $\mathbb{Q} = \left\{ \frac{a}{b} \mid a, b \in \mathbb{Z}, b \neq 0 \right\}$ the set of rational numbers and $\mathbb{R}$ the set of real numbers. It is clear that $-5 \in \mathbb{Z}, \frac{3}{5} \in \mathbb{Q}, \sqrt{23} \in \mathbb{R}, \pi \notin \mathbb{N}, \sqrt{23} \notin \mathbb{Z}$.

According to the number of elements of each set, the set can be **finite** or **infinite**. For example, if we consider the set of grains of sand in the world, although the amount is very large and possibly we could never finish counting, this set is finite. The numbers set are infinite. The set with no elements is called **empty set** [5–8]. This set is unique and it is represented by $\emptyset$. On the other hand, a finite set consisting of one element is called **singleton**.

Remark 1.1 $\emptyset$ and $\{\emptyset\}$ are totally different. The first one is a set without elements, whereas the another set is a singleton with the unique element that is $\emptyset$.

© Springer Nature Switzerland AG 2019
R. Martínez-Guerra et al., *Algebraic and Differential Methods for Nonlinear Control Theory*, Mathematical and Analytical Techniques with Applications to Engineering, https://doi.org/10.1007/978-3-030-12025-2_1

Definition 1.2 Let A, B two set. It is said that A is a **subset** of B, $A \subset B$ (or $B \supset A$), if and only if every element of A is element of B, i.e., $a \in A \Rightarrow a \in B$, $\forall a \in A$. (If $A \neq B$ we call A a proper subset of B).

According to this new concept, note that if $a \in A$ then we can to write an equivalent notation for this fact, i.e. $\{a\} \subset A$. Other examples of subsets are given by the following chain of inclusions: $\mathbb{N} \subset \mathbb{Z} \subset \mathbb{Q} \subset \mathbb{R} \subset \mathbb{C}$, the larger field, $\mathbb{C}$ is called an extension field of $\mathbb{R}$, similarly $\mathbb{R}$ is an extension field of $\mathbb{Q}$, this concept will have relevance in Chap. 10.

Remark 1.2 Let A an arbitrary set, then the empty set satisfies $\emptyset \subset A$ for all set A. To prove this fact, let us suppose that the statement is false. It is false only if $\emptyset$ has an element that does not belong to A. Since $\emptyset$ has not elements, then is not true and therefore $\emptyset \subset A$ for every A.

Remark 1.3 Let A, B, C any set, according to the Definition 1.2,

- $A \subset A$. This property states that the inclusion is **reflexive** (every set is a subset of itself).
- The inclusion is **antisymmetric**. This property is written as follows: $A \subset B$ and $B \subset A$ then $A = B$.
- If $A \subset B$ and $B \subset C$ then $A \subset C$, i.e., the inclusion is **transitive**.

The condition of equality between two sets is given from the following axiom.

Property 1.1 (Axiom of extension): *Two sets are equal if and only if they have the same elements.*

The axiom of extension states that if A and B are set such that $A \subset B$ and $B \subset A$ then A and B have the same elements and therefore $A = B$. This is the more general form to prove the equality between two set and this is the necessary and sufficient condition. Note that equality is **symmetric**, i.e., if $A = B$ then $B = A$. In particular, when $A = A$ it said that equality is **reflexive**. Finally, the relationship of equality is **transitive**. This property is analogous to the property of transitivity for subsets, i.e., $A = B$ and $B = C$ then $A = C$ [10–12].

1.1.1 Set Operations and Other Properties

Sometimes, when we have two set, A and B that are not necessarily equals, it is necessary to define a third set whose elements are given by the above sets. So that, it is necessary to define an operation in the sets with that feature. Before the definition is established, an important axiom that determines these new operations:

Property 1.2 (Axiom of unions): *For every collection of sets there exists a set that contains all the elements that belong to at least one set of the given collection.*

Although is complicated to define a set namely **universe**, that contains every set, in the next definition we will suppose that there exists a set (non empty) that contains a finite number of sets.

Definition 1.3 Let A, B two subsets of a set S. The **union** of two sets A and B is the set

$$A \cup B = \{s \in S \mid s \in A \text{ or } s \in B\}$$

where, we say $s \in A \cup B$ if and only if $s \in A$, $s \in B$ or s is in both sets.

The other important operation is the intersection between two sets, defined as follows.

Definition 1.4 Let A, B two subsets of a set S. The **intersection** of two sets A and B is the set

$$A \cap B = \{s \in S \mid s \in A \text{ and } s \in B\}$$

where, we say $s \in A \cap B$ if and only if $s \in A$ and $s \in B$.

Remark 1.4 These operations can be generalized for more sets. Let $\{A_\alpha\}_{\alpha \in I}$ a collection of subsets of S, then:

- $\bigcup_{\alpha \in I} A_\alpha = \{s \in S \mid s \in A_\alpha \text{ for some } \alpha \in I, \ I \text{ is an index set}\}$
- $\bigcap_{\alpha \in I} A_\alpha = \{s \in S \mid s \in A_\alpha \text{ for all } \alpha \in I, \ I \text{ is an index set}\}$

In any application of set theory, all sets that intervene can be considered as subsets of a fixed set. This set is called **universal set** (or simply universe) and it is denoted by $\mathcal{U}$.

Definition 1.5 Let the universe $\mathcal{U}$ and an arbitrary subset A of $\mathcal{U}$, the complement of set A respect to $\mathcal{U}$, to the set of all elements that do **not** belong to A but that belong to $\mathcal{U}$ [11, 13].

In analogous manner the union an intersection operations have some particular properties. We establish the following assertion. Note that each assertion should be demonstrated.

1. **Associativity**

   - $A \cup (B \cup C) = (A \cup B) \cup C$
   - $A \cap (B \cap C) = (A \cap B) \cap C$

2. **Commutativity**

   - $A \cup B = B \cup A$
   - $A \cap B = B \cap A$

3. **Distributivity**

   - $A \cup (B \cap C) = (A \cup B) \cap (A \cup C)$
   - $A \cap (B \cup C) = (A \cap B) \cup (A \cap C)$

4. **Idempotency**

   - $A \cup A = A$ for all set A.
   - $A \cap A = A$ for all set A.

5. **De Morgan's laws**

   - $(A \cup B)^c = A^c \cap B^c$
   - $(A \cap B)^c = A^c \cup B^c$

To prove each one of the claims, it is necessary to establish the equality between both sets. That is to say, we prove the two inclusions:

$$A = B \Leftrightarrow A \subset B \text{ and } B \subset A$$

For example, for the case of associativity of union, we have the following proof. Let $x \in A \cup (B \cup C)$, then

$$
\begin{aligned}
x \in A \cup (B \cup C) &\Leftrightarrow x \in A \text{ or } x \in (B \cup C) \\
&\Leftrightarrow x \in A \text{ or } x \in B \text{ or } x \in C \\
&\Leftrightarrow x \in (A \cup B) \text{ or } x \in C \\
&\Leftrightarrow x \in (A \cup B) \cup C
\end{aligned}
$$

Thus $A \cup (B \cup C) = (A \cup B) \cup C$.

At the end of this section, we will establish a proposition with more properties. Now, we establish the following definition.

Definition 1.6 Let A, B subsets of a set S. A and B are said to be **disjoint** if they have not elements in common, that is to say: $A \cap B = \emptyset$.

Note, that this definition can be extended to any family of sets: A family of sets is **pairwise disjoint** or **mutually disjoint** if every two different sets in the family are disjoint ($A_i, A_j \in \mathscr{A}, i, j \in I, A_i \cap A_j = \emptyset$).

Another important operation is the difference between two sets, defined as follows.

Definition 1.7 Let A, B subsets of a set S. The **difference** of the sets A and B is the set of all elements of set A that are not belong to the set B, i.e.,

$$A \setminus B = A - B = \{s \in S \mid s \in A, \ s \notin B\} = A \cap B^c$$

where B^c denotes the complement of B.

Example 1.2 Let $K = \{1, 2, 3, 4, 5\}$, $L = \{1, 2, 3, 4, 5, 6, 7, 8\}$, and $M = \{1, 2, 3, 4, 5, 6, 7, 8, 9, 10, 11, 12\}$. Note that $K \subset L \subset M$. According to the difference of sets, we have that

$$L - K = \{6, 7, 8\}$$
$$M - L = \{9, 10, 11, 12\}$$
$$M - K = \{6, 7, 8, 9, 10, 11, 12\}$$

Note as these sets are related by the union between them, and by their number of elements denominated cardinality of the set, i.e.:

$$M - K = (M - L) \cup (L - K)$$

and $7 = card(M - K) = card(M - L) + card(L - K) = 4 + 3$. This idea will be generalized and stated in Chap. 10 as a theorem for a tower of fields and their transcendence degree.

Remark 1.5 According to the above definition, the complement of a set A can be written as follows:

$$A^c = \{x \mid x \in \mathcal{U} \text{ and } x \notin A\} = \mathcal{U} \setminus A$$

that is, the difference of the universe with the set A.

The properties of difference of set are the following:

1. For every set A, $A \setminus A = \emptyset$.
2. For every set A, $A \setminus \emptyset = A$.
3. For every set A, $\emptyset \setminus A = \emptyset$.
4. Let A and B are nonempty sets. If $A - B = B - A$ then $A = B$

The latter property can be demonstrated as follows. By definition, we have

$$A - B = \{x \mid x \in A \text{ and } x \notin B\}$$
$$B - A = \{x \mid x \in B \text{ and } x \notin A\}$$

If $A - B = B - A$ then every element of $A \setminus B$ is an element of $B \setminus A$, but in the set $A \setminus B$ there are only elements of A. In the same manner, all element of $B \setminus A$ is an element of $A \setminus B$ but in the set $B \setminus A$ there are only elements of B, therefore, $A = B$.

1.2 Equivalence Relations

Definition 1.8 Let A a nonempty set. A **relation** in A is a subset R of $A \times A$.

Remark 1.6 We write $a \sim b$ if $(a, b) \in R$.

Definition 1.9 A relation R in A it is said to be:

1. **Reflexive**, if $a \sim a \; \forall a \in A$
2. **Symmetric**, if $a \sim b \Rightarrow b \sim a \; \forall a, b \in A$
3. **Transitive**, if $a \sim b$ and $b \sim c \Rightarrow a \sim c \; \forall a, b, c \in A$

Definition 1.10 A relation R in A it is said to be **equivalence relation** if R is reflexive, symmetric and transitive [14, 15].

Example 1.3 Let $A = \mathbb{Z}$. Let $a, b \in A$. It said to be a and b have the same parity, if a and b are both even or are both odd. We say that a and b are related, $a \sim b$, if a and b have the same parity. Indeed, this is a equivalence relation:

- $a \sim a \Leftrightarrow (-1)^a = (-1)^a$, $a - a = 0$ is even.
- $a \sim b \Leftrightarrow (-1)^a = (-1)^b \Leftrightarrow (-1)^b = (-1)^a \Leftrightarrow b \sim a$.
- $a \sim b \Leftrightarrow (-1)^a = (-1)^b$ and $b \sim c \Leftrightarrow (-1)^b = (-1)^c$. By virtue of the properties of equality, we have $(-1)^a = (-1)^b = (-1)^c$. Therefore $a \sim c$.

Definition 1.11 Let $m, n \in \mathbb{Z}$. It is said that m divides n if there exists $q \in \mathbb{Z}$ such that $n = mq$. In other words, m is a factor of n or n it is a **multiple** of m.

Remark 1.7 The notation for divisibility of numbers is the following:

- If m divides n, we write $m \mid n$.
- In other case, i.e., if m not divides n, we write $m \nmid n$.

Definition 1.12 Let $n \in \mathbb{N}$, $a, b \in \mathbb{Z}$. It said to be a is **congruent** with b **module** n if $n \mid (b - a)$. We write $a \equiv b \mod n$. The congruence module n is a equivalence relation.

Definition 1.13 Inclusion is not an equivalence relation. Note that $\mathbb{N} \subset \mathbb{Z}$ but $\mathbb{Z} \not\subset \mathbb{N}$ ($\subset$ is not symmetric).

Definition 1.14 Let $\sim$ an equivalence relation in A. The **equivalence class** of $a \in A$ is denoted by the following set:

$$\mathscr{C} = [a] = \{x \in A \mid x \sim a\}$$

The elements of $[a]$ are the elements of A that are related with a. If $\mathscr{C}$ is an equivalence class, then any element of $\mathscr{C}$ is a **representative** of $\mathscr{C}$.

Example 1.4 In the case of the parity we have:

- $[0] = \{\ldots, -4, -2, 0, 2, 4, \ldots\} = \{2m \mid m \in \mathbb{Z}\}$
- $[1] = \{\ldots, -5, -3, -1, 1, 3, 5, \ldots\} = \{2m + 1 \mid m \in \mathbb{Z}\}$

Example 1.5 According to $a \equiv b \mod n$, if $n = 5$, we have:

- $[0] = \{\ldots, -10, -5, 0, 5, 10, \ldots\} = \{5m \mid m \in \mathbb{Z}\}$
- $[1] = \{\ldots, -9, -4, 1, 6, 11, \ldots\} = \{5m + 1 \mid m \in \mathbb{Z}\}$

- $[2] = \{\ldots, -8, -3, 2, 7, 12, \ldots\} = \{5m + 2 \mid m \in \mathbb{Z}\}$
- $[3] = \{\ldots, -7, -2, 3, 8, 13, \ldots\} = \{5m + 3 \mid m \in \mathbb{Z}\}$
- $[4] = \{\ldots, -6, -1, 4, 9, 14, \ldots\} = \{5m + 4 \mid m \in \mathbb{Z}\}$

Theorem 1.1 *If $\sim$ is an equivalence relation in A, then*

$$A = \bigcup_{a \in A} [a]$$

In addition, if $[a] \neq [b] \Rightarrow [a] \cap [b] = \emptyset$.

Proof First, we prove the double contention of subsets.

1. Let $a \in A$. Then $a \in [a]$ and therefore $a \in \bigcup_{a \in A} [a]$. This implies that $A \subset \bigcup_{a \in A} [a]$.
2. Given that $[a] \subset A \ \forall a \in A$, therefore $\bigcup_{a \in A} [a] \subset A$.

Hence

$$A = \bigcup_{a \in A} [a]$$

Finally we will use the contrapositive form to prove the last statement. Suppose that $[a] \cap [b] \neq \emptyset$. Let $c \in [a] \cap [b]$, then we have $c \sim a$ and $c \sim b$, therefore $a \sim b$. Now, if $x \in [a]$, then $x \sim a$ and $x \sim b$ (since $a \sim b$) and hence $x \in [b]$. This implies $[a] \subset [b]$ and in a similar manner, $[b] \subset [a]$. Therefore $[a] = [b]$. $\square$

Definition 1.15 Let A a nonempty set. A **partition** of A, is a collection $\{A_\alpha\}_{\alpha \in I}$ of subsets of A such that:

$$A = \bigcup_{\alpha \in I} A_\alpha$$

In addition, if $\alpha \neq \beta$ then $A_\alpha \cap A_\beta = \emptyset$. A_α are so called **elements of the partition**.

Remark 1.8 Let A a nonempty set. An equivalence relation in A gives rise to a partition of A. The elements of this partition, are the equivalence classes. Reciprocally, a partition of A generates an equivalence relation. In conclusion, let $a, b \in A$. a and b are said to be related if and only if they are in the same element of the partition.

Definition 1.16 Let A a nonempty set and R (or $\sim$) an equivalence relation in A. A is said to be a **quotient set** denoted by:

$$A/R = A/\sim = \{[a] \mid a \in A\}$$

Example 1.6 Let $n \in \mathbb{N}$. Consider the congruence module n in $\mathbb{Z}$ (that is an equivalence relation). The quotient set of $\mathbb{Z}$ is given by

$$\mathbb{Z}/n\mathbb{Z} = \{[0], \ldots, [n-1]\}$$

Sometimes, $\mathbb{Z}/n\mathbb{Z}$ is denoted by $\mathbb{Z}/n$ or $\mathbb{Z}_n$. As particular cases, we have:

1. $\mathbb{Z}/2\mathbb{Z} = \{[0], [1]\}$
2. $\mathbb{Z}/5\mathbb{Z} = \{[0], [1], [2], [3], [4]\}$

1.3 Functions or Maps

Definition 1.17 Let A, B nonempty sets. A **function** or **map** is a relation given by a set A and a set B with the following property: each element $a \in A$ is related to exactly one and only one element $b \in B$. In other words, a function or map is a rule that assigns to each element a of A [16, 17], one and only one element b of B. If $f : A \to B$ is a function, the set A is called **domain** of the function and the set B is the **codomain** or range of f. A function f is commonly denoted by:

$$f : A \to B$$
$$a \longmapsto b$$

The notation $f(a) = b$ is equivalent to say $a \longmapsto b$. The **image of** $a \in A$ is $f(a) \in B$. The **image of the function** f is:

$$\mathrm{Im}\, f = \{f(a) \mid a \in A\} \subset B$$

Example 1.7   1. Let f defined by:

$$f : \mathbb{R} \to \mathbb{R}$$
$$a \longmapsto a^2$$

It is not hard to see that some elements are given by:

$$f(2) = 4, \ f(-2) = 4, \ f(5) = 25, \ f(-5) = 25, \ f(0) = 0.$$

2. Let g defined by:

$$g : \mathbb{Z} \to \mathbb{Z}$$
$$n \longmapsto 2n + 1$$

Some elements using the correspondence rule, are the following:

$$g(2) = 5, g(-2) = -3, g(5) = 11, g(-5) = -9, g(0) = 1.$$

3. Let A and B nonempty sets. The functions

$$\pi_1 : A \times B \to A$$
$$(a, b) \longmapsto a$$

and

$$\pi_2 : A \times B \rightarrow B$$
$$(a, b) \longmapsto b$$

are called **projections**.
4. Let A and B nonempty sets. Let $b_0 \in B$. The **constant function** is defined by

$$f : A \rightarrow B$$
$$a \longmapsto b_0$$

In other words, $f(a) = b_0 \ \forall a \in A$. In particular, when $B = A$, we have the **identity function**:

$$\mathrm{id}_A : A \rightarrow A$$
$$a \longmapsto a$$

Definition 1.18 The functions f and g are **equal**, if they have the same domain and range, i.e., $f = g$ if and only if $f : A \rightarrow B$ and $g : B \rightarrow A$. In addition,

$$f(a) = g(a) \ \forall a \in A$$

Example 1.8 Let f and g the functions defined by

$$f : \mathbb{Z} \rightarrow \mathbb{Z} \qquad\qquad g : \mathbb{Z} \rightarrow \mathbb{Z}$$
$$a \longmapsto a^2 - 1 \quad , \qquad a \longmapsto (a+1)(a-1)$$

Note that domain and codomain of both functions are equal, i.e., $\mathbb{Z}$. Now, we choose an arbitrary element $z \in \mathbb{Z}$, then $f(z) = z^2 - 1 = (z+1)(z-1) = g(z) \ \forall z \in \mathbb{Z}$. Therefore $f = g$.

1.3.1 Classification of Functions or Maps

Definition 1.19 The function $f : A \rightarrow B$ is called **injective function** if $a_1 \neq a_2$ then $f(a_1) \neq f(a_2) \ \forall a_1, a_2 \in A$. Equivalently if $f(a_1) = f(a_2)$ then $a_1 = a_2 \ \forall a_1, a_2 \in A$.

Example 1.9 Let $g : \mathbb{R} \setminus \{-2\} \rightarrow \mathbb{R}$ given by

$$g(x) = \frac{2x - 1}{x + 2} \tag{1.1}$$

Let $x_1, x_2 \in D_g$. Suppose that $g(x_1) = g(x_2)$ then

$$\frac{2x_1 - 1}{x_1 + 2} = \frac{2x_2 - 1}{x_2 + 2}$$

if and only if

$$(2x_1 - 1)(x_2 + 2) = (2x_2 - 1)(x_1 + 2)$$

this yields to $x_1 = x_2$.

Hence g is injective.

Definition 1.20 A **surjective function** or **onto** is a function whose image is equal to its codomain. Equivalently, a function $f : A \to B$ with domain A and codomain B is surjective if for every $b \in B$ there exists at least one $a \in A$ such that $f(a) = b$. In other words, $f : A \to B$ is onto if $\mathrm{Im} f = B$ [18, 19].

Example 1.10 Let $g : \mathbb{R} \setminus \{-2\} \to \mathbb{R}$ given by

$$g(x) = \frac{2x - 1}{x + 2}$$

This function is surjective since for all $y \in \mathbb{R} \setminus \{-2\}$ exists $x = \frac{1+2y}{2-y}$ such that

$$g(x) = \frac{2\left(\frac{1+2y}{2-y}\right) - 1}{\left(\frac{1+2y}{2-y}\right) + 2} = y$$

Definition 1.21 The function $f : A \to B$ is called **bijective function** or **bijection** if f is an injective and surjective function.

Example 1.11 The function g defined above is injective and surjective, thus is bijective.

Definition 1.22 Let $f : A \to B$ a bijective function. In this case, the **inverse function** of f is:

$$f^{-1} : B \to A$$

The inverse function is given by $f^{-1}(b) = a$ if and only if $f(a) = b$.

Example 1.12 The function g defined by (1.1) is a bijective function then the inverse function g^{-1} exists given by

$$g^{-1}(x) = \frac{2x + 1}{2 - x}$$

Proposition 1.1 *Let $f : A \to B$. If f is bijective, then its inverse function $f^{-1} : B \to A$ is also bijective and $\left(f^{-1}\right)^{-1} = f$.*

Proof The proof is left to the reader. □

Definition 1.23 Let $g : A \rightarrow B$ and $f : B \rightarrow C$ functions. The **composition** of f and g, denoted by $f \circ g$ is the function:

$$f \circ g : A \rightarrow C$$

and is given by $(f \circ g)(a) = f(g(a))$.

Example 1.13 $f : \mathbb{R} \rightarrow \mathbb{R}$, $g : \mathbb{R} \rightarrow \mathbb{R}$ defined by: $f(x) = x^2 - 1$, $g(x) = \sin(x)$, hence $(f \circ g)(x) = -\cos^2(x)$.

Example 1.14 In general $f \circ g \neq g \circ f$. Let f and g the functions defined by

$$f : \mathbb{R} \rightarrow \mathbb{R} \qquad g : \mathbb{R} \rightarrow \mathbb{R}$$
$$a \longmapsto a^2 \qquad , \qquad a \longmapsto (a + 1)$$

From the definition, we have:

$$f \circ g : \mathbb{R} \rightarrow \mathbb{R} \qquad g \circ f : \mathbb{R} \rightarrow \mathbb{R}$$
$$a \longmapsto (a + 1)^2 \qquad , \qquad a \longmapsto a^2 + 1$$

Therefore, $f \circ g \neq g \circ f$.

Proposition 1.2 *Let* $h : A \rightarrow B$, $g : B \rightarrow C$, $f : C \rightarrow D$ *functions. Then:*

$$(f \circ g) \circ h = f \circ (g \circ h)$$

Proof We have $(f \circ g) \circ h : A \rightarrow D$ and $f \circ (g \circ h) : A \rightarrow D$, so the domain and the codomain are the same for both functions. Let $a \in A$, then we have the following equalities:

$$((f \circ g) \circ h)(a) = (f \circ g)(h(a)) = f(g(h(a))) = f((g \circ h)(a)) = (f \circ (g \circ h))(a)$$

This yields to
$$((f \circ g) \circ h)(a) = (f \circ (g \circ h))(a) \ \forall a \in A$$

Finally, we conclude that $(f \circ g) \circ h = f \circ (g \circ h)$. □

Proposition 1.3 *Let* $g : A \rightarrow B$, $f : B \rightarrow C$ *functions. Then:*

1. *If* f, g *are injective functions, then* $f \circ g$ *is injective.*
2. *If* f, g *are surjective functions, then* $f \circ g$ *is surjective.*
3. *If* f, g *are bijective functions, then* $f \circ g$ *is bijective.*

Proof The proof is left to the reader.   □

Proposition 1.4 *Let* $g : A \to B$, $f : B \to C$ *functions. Then:*

1. *If* $f \circ g$ *is injective then* g *is injective.*
2. *If* $f \circ g$ *is injective then* f *is not necessarily injective.*
3. *If* $f \circ g$ *is surjective then* f *is surjective.*
4. *If* $f \circ g$ *is surjective then* g *is not necessarily surjective.*
5. *If* $f \circ g$ *is bijective then* g *is injective and* f *is surjective.*
6. *If* $f \circ g$ *is bijective then* f *and* g *are not necessarily injective and surjective respectively.*

Proof The proof is left to the reader.   □

Proposition 1.5 *Let* $f : A \to B$ *a bijective function and* $f^{-1} : B \to A$ *the inverse of* f, *then:*

1. $f \circ f^{-1} = id_B$
2. $f^{-1} \circ f = id_A$

Proof First, note that $f \circ f^{-1} : B \to B$, $id_B : B \to B$, $f^{-1} \circ f : A \to A$ and $id_A : A \to A$. Now, we consider the first case. Let $b \in B$. As the function f is bijective, then there exists a unique $a \in A$ such that $f(a) = b$. According to the definition of inverse function, $f^{-1}(b) = a$, then we have:

$$(f \circ f^{-1})(b) = f(f^{-1}(b)) = f(a) = b = id_B(b)$$

Therefore $f \circ f^{-1} = id_B$. For the second case, let $a \in A$ and $b = f(a)$. Since the function f is bijective, then $f^{-1}(b) = a$. Therefore:

$$(f^{-1} \circ f)(a) = f^{-1}(f(a)) = f^{-1}(b) = a = id_A(a)$$

Thus $f^{-1} \circ f = id_A$.   □

Proposition 1.6 *If* $f : A \to B$ *is function and there exists a function* $g : B \to A$ *such that* $g \circ f = id_A$ *and* $f \circ g = id_B$, *then* f *is bijective and* $f^{-1} = g$.

Proof It is trivial considering the Proposition 1.4. According to this result, we refer to the bijective function as an **invertible function**.   □

Proposition 1.7 *Let* $f : A \to B$ *a function. Then*

1. $f \circ id_A = f$
2. $id_B \circ f = f$

Proof Note that $f \circ id_A : A \to B$ and $id_B \circ f : A \to B$. Let $a \in A$ then $(f \circ id_A)$ $(a) = f(id_A(a)) = f(a)$. In the same manner $(id_B \circ f)(a) = id_B(f(a)) = f(a)$. Therefore $f \circ id_A = f$ and $id_B \circ f = f$.   □

Proposition 1.8 *Let* $g : A \to B$ *and* $f : B \to C$ *bijective functions. Then*

$$(f \circ g)^{-1} = g^{-1} \circ f^{-1}$$

Proof From the Proposition 1.3, $f \circ g : A \to C$ is bijective and $(f \circ g)^{-1} : C \to A$. Since f and g are bijective functions then $g^{-1} : B \to A$ and $f^{-1} : C \to B$. Then $g^{-1} \circ f^{-1} : C \to A$, therefore $(f \circ g)^{-1}$ and $g^{-1} \circ f^{-1}$ have the same domain and codomain. Let $c \in C$, then there exists $a \in A$ and $b \in B$ such that $g(a) = b$ and $f(b) = c$. Then we have $(f \circ g)(a) = f(g(a)) = f(b) = c$, hence $(f \circ g)^{-1}(c) = a$. On the other hand, $f^{-1}(c) = b$ and $g^{-1}(b) = a$, then $(g^{-1} \circ f^{-1})(c) = g^{-1}(f^{-1}(c)) = g^{-1}(b) = a$. Therefore $(f \circ g)^{-1} = g^{-1} \circ f^{-1}$. $\qquad\square$

1.4 Well-Ordering Principle and Mathematical Induction

Consider the well-ordering principle axiom that is stated as follows. **Well-ordering principle**: If A is subset nonempty of $\mathbb{N}$, then there exists $m \in A$ such that $m \leq a \; \forall a \in A$, i.e., m is the least element of A.

Theorem 1.2 (Mathematical induction principle) *Let* $P(n)$ *a statement about natural numbers* $\mathbb{N}$ *such that:*

1. $P(1)$ *is valid.*
2. *Suppose that statement is true for* $n = k$, *i.e.,* $P(k)$ *is valid, then* $P(k + 1)$ *is valid* $\forall k \in \mathbb{N}$.

then $P(n)$ *is true for all* $n \in \mathbb{N}$.

Proof By contradiction. Suppose that the theorem is false, then there exists a statement $P(n)$ such that satisfies (1) and (2) but $P(n)$ is not valid for all $n \in \mathbb{N}$. By using the well-ordering principle, there exists a minimum natural number m such that $P(m)$ is not valid. Since (2) is true, then $m > 1$. By using the minimality of m we have $P(m - 1)$ is valid. According to (2) then $P(m)$ is valid, but this contradicts the fact that $P(m)$ is not valid. Hence $P(n)$ is valid $\forall n \in \mathbb{N}$. $\qquad\square$

Remark 1.9 Some comments about the mathematical induction principle, then we can do some the following modifications.

- Change 1 for any $n_0 \in \mathbb{Z}^+$. Conclude that $P(n)$ is valid $\forall n \geq n_0$.
- Suppose that $P(1)$ is valid and $P(j)$ is valid too $\forall j < k$, then $P(k)$ is valid $\forall k \in \mathbb{N}$. Conclude that $P(n)$ is valid $\forall n \in \mathbb{N}$.
- Suppose that $P(n_0)$ is valid $\forall j < k$ then $P(k)$ is valid $\forall k \geq n_0$. Conclude that $P(n)$ is valid $\forall n \geq n_0$.

Example 1.15 Show that the sum of the first n natural numbers is:

$$1 + 2 + 3 + \cdots + n = \frac{n(n + 1)}{2} \quad \forall n \in \mathbb{N} \tag{1.2}$$

Proof Using induction over n. Let S_n the set of positive integers n such that satisfy (1.2), i.e,

$$S_n = 1 + 2 + 3 + \cdots + n$$

First we verify that 1 satisfies S_n:

$$S_1 = \frac{1(1+1)}{2} = 1$$

Let us prove the induction step. Suppose that k satisfies S_n (induction hypothesis), i.e.

$$S_k = 1 + 2 + 3 + \cdots + k = \frac{k(k+1)}{2} \tag{1.3}$$

and prove that

$$S_{k+1} = \frac{(k+1)(k+2)}{2} \tag{1.4}$$

Indeed,

$$S_{k+1} = S_k + (k+1) = 1 + 2 + 3 + \cdots + k + (k+1) = \frac{k(k+1)}{2} + (k+1) = \frac{(k+1)(k+2)}{2}$$

Therefore:

$$1 + 2 + 3 + \cdots + n = \frac{n(n+1)}{2} \quad \forall n \in \mathbb{N} \qquad \square$$

Remark 1.10 In another way, let $S_1(n) = 1 + 2 + 3 + \cdots + n$. We write the sum $S_1(n)$ twice with the second one written in opposite order:

$$S_1(n) = 1 + 2 + 3 + \cdots + n$$
$$S_1(n) = n + (n-1) + (n-2) + \cdots + 2 + 1$$

Then we add the two expressions:

$$2S_1(n) = (n+1) + (n+1) + (n+1) + (n+1) + \cdots + (n+1) + (n+1)$$

That is, $2S_1(n) = n(n+1)$ and therefore

$$S_1(n) = 1 + 2 + 3 + \cdots + n = \frac{n(n+1)}{2} \quad \forall n \in \mathbb{N}$$

Example 1.16 Prove the following formula by induction.

$$1^2 + 2^2 + \cdots + n^2 = \frac{n(n+1)(2n+1)}{6} \tag{1.5}$$

Proof Denote by S_n the sum:

$$S_n = 1^2 + 2^2 + \cdots + n^2$$

Let A the set of positive integers such that satisfies (1.5). Firstly is verified that $1 \in A$:

$$1^2 = 1 = \frac{6}{6} = \frac{(1)(2)(3)}{6} = \frac{(1)(1+1)\,[2(1)+1]}{6}$$

Now the induction step is tested. Suppose that $k \in A$ (induction hypothesis) then

$$S_k = \frac{k(k+1)(2k+1)}{6} \tag{1.6}$$

Let us show that

$$S_{k+1} = \frac{(k+1)[(k+1)+1][2(k+1)+1]}{6}$$

Indeed, adding $(k+1)^2$ on both sides of (1.6) and making algebraic manipulations:

$$
\begin{aligned}
S_{k+1} &= S_k + (k+1)^2 \\
&= \frac{k(k+1)(2k+1)}{6} + (k+1)^2 \\
&= \frac{k(k+1)(2k+1) + 6(k+1)^2}{6} \\
&= \frac{(k+1)\,[k(2k+1) + 6(k+1)]}{6} \\
&= \frac{(k+1)\,[2k^2 + 7k + 6]}{6} \\
&= \frac{(k+1)(k+2)(2k+3)}{6} \\
&= \frac{(k+1)\,[(k+1)+1]\,[2(k+1)+1]}{6}
\end{aligned}
$$

We have shown that $k+1 \in A$. Given that $1 \in A$ and for every positive integer k belonging to the set A, the number $k+1$ also belongs to the set A, by using the mathematical induction principle $A = \mathbb{N}$ and (1.5) is valid for every natural number n. □

Example 1.17

$$1^3 + 2^3 + \cdots + n^3 = (1 + \cdots + n)^2 \tag{1.7}$$

Proof We proceed by using mathematical induction on n. Let B the set of positive integers such that satisfies (1.7). Firstly we have

$$1^3 = 1 = 1^2$$

then $1 \in B$. Suppose that (1.7) is valid for $n = k$:

$$1^3 + 2^3 + \cdots + k^3 = (1 + \cdots + k)^2 \tag{1.8}$$

Equation above represents the induction hypothesis. Let us demonstrate that the hypothesis is valid for $n = k + 1$, that is

$$1^3 + 2^3 + \cdots + k^3 + (k+1)^3 = [1 + \cdots + k + (k+1)]^2$$

Then

$$
\begin{aligned}
[1 + \cdots + k + (k+1)]^2 &= (1 + \cdots + k)^2 + 2(1 + \cdots + k)(k+1) + (k+1)^2 \\
&= 1^3 + 2^3 + \cdots + k^3 + 2\left[\frac{k(k+1)}{2}\right](k+1) + (k+1)^2 \\
&= 1^3 + 2^3 + \cdots + k^3 + k(k+1)^2 + (k+1)^2 \\
&= 1^3 + 2^3 + \cdots + k^3 + (k+1)^2(k+1) \\
&= 1^3 + 2^3 + \cdots + k^3 + (k+1)^3
\end{aligned}
$$

therefore $k + 1 \in B$ which means that $B = \mathbb{N}$. Based in the mathematical induction principle, (1.7) is valid for every natural number n. □

References

1. Adamson, I. T.: A Set Theory Workbook. Springer Science & Business Median, New York (2012)
2. Hammack, R.: Book of proof. Virginia Commonwealth University Mathematics (2013)
3. Lang, S.: Algebra. Springer, New York (2002)
4. Lipschutz, S.: Schaum's outline of theory and problems of set theory and related topics (1964)
5. Bourbaki, N.: Theory of Sets. Springer, Heidelberg (2004)
6. Halmos, P.R.: Naive Set Theory. Dover Publications, New York (2017)
7. Henkin, L.: On mathematical induction. Am. Math. Monthly **67**(4), 323–338 (1960)
8. Hrbacek, K., Jech, T.: Introduction to Set Theory, Revised and Expanded. CRC Press, Boca Raton (1999)
9. Sominskii, I.S.: The Method of Mathematical Induction (Popular Lectures in Mathematics). Pergamon Press, Oxford (1961)
10. Bourbaki, N. Commutative Algebra (Vol. 8), Hermann, Paris (1972)
11. Kuratowski, K.: Introduction to Set Theory and Topology. Elsevier, Amsterdam (2014)
12. Smith, D., Eggen, M., Andre, R.S.: A Transition to Advanced Mathematics. Nelson Education (2014)

13. Jech, T. J.: The axiom of choice. Courier Corporation (2008)
14. Kunz, E.: Introduction to Commutative Algebra and Algebraic Geometry. Birkhuser, Boston (1985)
15. Zariski, O., Samuel, P.: Commutative Algebra, vol. 1. Springer Science & Business Media, New York (1958)
16. Birkhoff, G., Mac Lane, S.: A Survey of Modern Algebra. Universities Press (1965)
17. Sontag, E.D.: Linear systems over commutative rings. A survey. Ricerche di Automatica **7**(1), 1–34 (1976)
18. Samuel, P., Serge, B.: Mthodes d'algbre abstraite en gomtrie algbrique. Springer, Heidelberg (1967)
19. Zariski, O., Samuel, P.: Commutative Algebra, vol. 2. Springer Science & Business Media, New York (1960)

Chapter 2
Group Theory

Abstract This chapter provides an introduction to group theory, the chapter begins with basic definitions about groups, then continues to introduce subgroups and its characteristics, followed by homomorphisms, the chapter concludes with various theorems about isomorphisms, along with the concepts, various examples are provided to facilitate the understanding of the theory. This material ca ben found in any introductory book on abstract algebra (also called modern algebra).

2.1 Basic Definitions

Definition 2.1 A **group** is a nonempty set G ($G \neq \emptyset$) together with an operation $*$ (called the group law of G) that combines any two elements a and b to form another element, denoted $a * b$ or ab or explicitly [1–5]:

$$* : G \times G \to G$$
$$(a, b) \mapsto a * b$$

To qualify as a group, the set and operation, $(G, *)$, must to satisfy four requirements known as the group axioms [6–8]:

1. $a * b \in G \ \forall a, b \in G$ (Closure).
2. $a * (b * c) = (a * b) * c \ \forall a, b, c \in G$ (Associativity).
3. There exists $e \in G$ such that $a * e = e * a = a \ \forall a \in G$ (Existence of identical element. e is the identity of G).
4. $\forall a \in G$ there exists $b \in G$ such that $a * b = b * a = e$ (In this case, $b = a^{-1}$ is called the inverse of a).

Remark 2.1 If a $(G, *)$ satisfies 1 and 2, is called semigroup. The semigroup together with 3 is called **monoid**.

Example 2.1 1. The set of integer numbers $\mathbb{Z}$ with the operation sum $+$, denoted by $(\mathbb{Z}, +)$, form a group.

© Springer Nature Switzerland AG 2019
R. Martínez-Guerra et al., *Algebraic and Differential Methods for Nonlinear Control Theory*, Mathematical and Analytical Techniques with Applications to Engineering, https://doi.org/10.1007/978-3-030-12025-2_2

2. The set of rational numbers $\mathbb{Q}$ with the operation sum $+$, denoted by $(\mathbb{Q}, +)$, form a group.
3. Let us consider the set of rational numbers $\mathbb{Q}$ without the zero, denoted by $\mathbb{Q}^* = \mathbb{Q} \setminus \{0\}$ with the operation multiplication $\cdot$, form a group $(\mathbb{Q}^*, \cdot)$.
4. Let $G = \{e\}$ and the operation $*$ defined as $e * e = e \in G$. The pair $(G, *)$, form the most basic group.
5. Let $G = \{e, a\}$ and the operation $*$ defined as

$$
\begin{array}{c|cc}
* & e & a \\
\hline
e & e & a \\
a & a & e
\end{array}
$$

6. The set of matrix $G(n, \mathbb{R})$ (matrices of $n \times n$ with coefficients in $\mathbb{R}$) is a group non-commutative, i.e., $AB \neq BA$.
7. The set of matrices defined as

$$
\begin{bmatrix}
\cos \theta & \sin \theta \\
\sin \theta & \cos \theta
\end{bmatrix}
$$

is a group commutative with the matrix product.

Definition 2.2 Let G a group. It is said that G is an Abelian group [7, 9] if

$$
a * b = b * a \;\; \forall a, b \in G
$$

Definition 2.3 Let G a group. The order of a group, denoted by $| G |$, consists in the number of elements of G.

Note that if G is a countable set of elements, then the order of G is finite. In other case, G is called a infinite group, for example $|\{e\}| = 1$ and $|\mathbb{R}| = \infty$.

Example 2.2 1. The set $\mathbb{Z}$ with the operation $\cdot$ usual product is not a group because there is not a multiplicative inverse.
2. Let $\mathbb{Z}^* = \mathbb{Z} \cup \{0\}$ with the operation $\cdot$ (usual product), is $\mathbb{Z}^*$ a group? The proof is left to the reader as an exercise.
3. Let $G = \mathbb{R} \setminus \{0\}$. If we define the operation $a * b = a^2 b$, is G a group? The proof is left to the reader as an exercise.
4. The equivalence class $\mathbb{Z}/n\mathbb{Z}$ with the operation $[a] + [b] = [a + b]$ is a group, where $[a] = \{x \in \mathbb{Z} \mid a \sim x, a \cong x \mod n, n \mid (a - x)\}$. The proof is left to the reader as an exercise.

Proposition 2.1 *Let G a group, then:*

1. *The identity element is unique.*
2. $\forall a \in G \; a^{-1}$ *is unique.*
3. $\forall a \in G, \left(a^{-1}\right)^{-1} = a.$
4. $\forall a, b \in G, (ab)^{-1} = b^{-1}a^{-1}.$

5. In general, $(a_1 a_2 \cdots a_{n-1} a_n)^{-1} = a_n^{-1} a_{n-1}^{-1} \cdots a_2^{-1} a_1^{-1}$

Proof 1. Let e_1, e_2 identity elements of G, with $e_1 \neq e_2$. If e_1 is an identity element, then $e_1 e_2 = e_2 e_1 = e_2$. If e_2 is an identity element, then $e_2 e_1 = e_1 e_2 = e_1$, so we conclude that $e_1 = e_2$ i.e., the identity element e is unique.

2. Let b, c inverse elements of a in G, with $b \neq c$. Note that $ab = ba = e$ and $ac = ca = e$ then $b = be = b(ac) = (ba)c = ec = c$, therefore $b = c$, i.e., a^{-1} is unique.

3. Note that $aa^{-1} = a^{-1}a = e$, then a is the inverse element of a^{-1}. Therefore $\left(a^{-1}\right)^{-1} = a$.

4. First applying the invertibility on the left, i.e., $(b^{-1}a^{-1})(ab) = b^{-1}(a^{-1}a)b = b^{-1}(eb) = b^{-1}b = e$. Now, applying the invertibility on the right: $(ab)(b^{-1}a^{-1}) = a(bb^{-1})a^{-1} = a(ea^{-1}) = aa^{-1} = e$. Hence $(ab)^{-1} = b^{-1}a^{-1}$.

5. By induction. Let $n = 2$ then $(a_1 a_2)^{-1} = a_2^{-1} a_1^{-1}$, so the result is valid for $n = 2$. Assume, for $n = k$ that $(a_1 a_2 \cdots a_{k-1} a_k)^{-1} = a_k^{-1} a_{k-1}^{-1} \cdots a_2^{-1} a_1^{-1}$ holds. It easy to verify that $a_{k+1}^{-1}(a_1 a_2 \cdots a_{k-1} a_k)^{-1} = a_{k+1}^{-1} a_k^{-1} a_{k-1}^{-1} \cdots a_2^{-1} a_1^{-1}$, hence, we have $(a_1 a_2 \cdots a_k a_{k+1})^{-1} = a_{k+1}^{-1} a_k^{-1} \cdots a_2^{-1} a_1^{-1}$. $\qquad\square$

Proposition 2.2 (Cancelation laws) *Let G a group. Then,*

1. *If $ab = ac$ then $b = c$ $\forall a, b, c \in G$.*
2. *If $ba = ca$ then $b = c$ $\forall a, b, c \in G$.*

Proof We have:

1. $b = eb = (a^{-1}a)b = a^{-1}(ab) = a^{-1}(ac) = (a^{-1}a)c = ec = c$
2. $b = be = b(aa^{-1}) = (ba)a^{-1} = (ca)a^{-1} = c(aa^{-1}) = ce = c$

Hence, 1. and 2. have been proved. $\qquad\square$

2.2 Subgroups

Definition 2.4 Let G a group and H (a nonempty set) such that $H \subset G$. It said that H is a subgroup of G if H is a group with respect to the operation of G. If H is a subgroup of G, it is denoted by $H < G$ [10–13].

Remark 2.2 Every group G has two trivial subgroups: G and $\{e\}$.

Proposition 2.3 *A nonempty subset $H \subset G$ is a subgroup of G if and only if H is closed with the operation of G and if $a \in H$ then $a^{-1} \in H$.*

Proof • $\Rightarrow$ (Sufficiency). Let H a subgroup of G by the definition of subgroup. In particular, H is a group, H is closed and there exists the inverse.
• $\Leftarrow$ (Necessity). H is closed, is nonempty and $a^{-1} \in H$ (closure of H). In addition, let $a, b, c \in H$, then $a(bc) = (ab)c$ because $H \subset G$. Therefore H is a subgroup. $\qquad\square$

Exercise 2.1   1. Let $G = \mathbb{Z}$ considering the usual sum. Let H the set of even numbers, i.e. $H = \{n \mid n = 2k, k \in \mathbb{Z}\}$. Is H a subgroup of G?
2. Let X a nonempty set. We consider $G = S_X$. Let $a \in X$ then $H(a) = \{f \in S_X \mid f(a) = a\}$. Prove that $H < G$ through function composition.
3. Let $G = \mathbb{C}^* = \mathbb{C} \setminus \{0\}$ and the usual product. Let $\mathscr{U} = \{z \in \mathbb{C}^* : \mid z \mid = 1\}$. Prove that $\mathscr{U} < G$.
4. Let G a group and $a \in G$. Let $C(a) = \{g \in G \mid ga = ag\}$. Prove that $C(a)$ is a subgroup of G.

The proof of the exercises 1–3 are left to the reader. The proof of exercise 4 is as follows. $C(a)$ is a group and in particular is a subgroup of G. Indeed. Firstly, $C(a) \neq \emptyset$ since $a = ea = ae$ then $e \in C(a)$. On the other hand, let $g_1, g_2 \in C(a)$, we have to verify that $g_1 g_2 \in C(a)$ (closure). Since $g_1, g_2 \in C(a)$, $g_1 a = a g_1, g_2 a = a g_2$ then $(g_1 g_2)a = g_1(g_2 a) = g_1(a g_2) = (g_1 a)g_2 = (a g_1)g_2 = a(g_1 g_2)$. Hence $g_1 g_2 \in C(a)$. Also if $g \in C(a)$ then $g^{-1} \in C(a)$, i.e., $g^{-1}a = ag^{-1}$. Since $a = ae = a(gg^{-1}) = (ag)g^{-1} = (ga)g^{-1}$ then $g^{-1}(a) = g^{-1}((ga)g^{-1}) = g^{-1}gag^{-1} = eag^{-1} = ag^{-1}$ then $g^{-1} \in C(a)$. Thus $C(a) < G$.

Definition 2.5 Let G a group and $a \in G$. The set

$$A = \langle a \rangle = \{a^i \mid i \in \mathbb{Z}\}$$

is a subgroup of G (A is nonempty since $a^0 = e \in A$. The usual product for $a^i, a^j \in A$ is given by $a^i a^j = a^{i+j}, i, j \in \mathbb{Z}, i + j \in \mathbb{Z}, a^{i+j} \in A$. On the other hand $a^i \in A$ then $a^{-i} = (a^i)^{-1} = (a^{-1})^i \in A$). The set $A = \langle a \rangle$ is called cyclic subgroup of G generated by a [14, 15].

Definition 2.6 Let G a group. It is said that G is cyclic if $G = \langle a \rangle$ for some $a \in G$.

Exercise 2.2   1. Let $(G_1, *), (G_2, \circ)$ and $(G_3, \triangle)$ groups, where $G_1 = \{e\}, G_2 = \{e, a\}, G_3 = \{e, a, b\}$ and each group operation is defined by the following tables:

$*$	e
e	e

$\circ$	e	a
e	e	a
a	a	e

$\triangle$	e	a	b
e	e	a	b
a	a	b	e
b	b	e	a

What is the generator element (or elements) in each group?
2. What is the generator element in the group defined by $\mathbb{Z}/2\mathbb{Z} = \mathbb{Z}_2 = \{[0], [1]\}$ with the usual sum?
3. Let G a group, such that $x^2 = e \, \forall x \in G$. Prove that G is an abelian group. Hint: Verify that $x = x^{-1} \forall x \in G$.

Definition 2.7 Let G a group, H a subgroup of G. For $a, b \in G$, it is said that a and b are congruent modulo H ($a \cong b \mod H$), if $ab^{-1} \in H$.

Exercise 2.3 According to above definition, prove that $\cong$ is a equivalence relation.

Definition 2.8 Let H a subgroup of G, $a \in G$. The set

$$Ha = \{ha \mid h \in H\}$$

is called right lateral class of H in G.

Lemma 2.1 *Let G a group, and $a \in G$, then*

$$Ha = \{x \in G \mid a \cong x \mod H\}$$

Proof Let a set defined by

$$[a] = \{x \in G \mid a \cong x \mod H\}$$

- (Sufficiency). Let $h \in H$ and $ha \in Ha$. We take an arbitrary element in Ha and we show that this element is in $[a]$, i.e., if $a(ha)^{-1} \in H$ then $a \cong ha \mod H$ and therefore $ha \in [a]$. Note that $a(ha)^{-1} = a(a^{-1}h^{-1}) = (aa^{-1})h^{-1} = eh^{-1} = h^{-1} \in H$, then $ha \in [a]$. So, we conclude that $Ha \subset [a]$.
- (Necessity). Let $x \in [a]$ then $a \cong x \mod H$, i.e., $ax^{-1} \in H$. In particular $(ax^{-1})^{-1} = xa^{-1} \in H$. Furthermore, let $h = xa^{-1} \in H$, then $ha = (xa^{-1})a = x(a^{-1}a) = xe = x \in Ha$, then $[a] \subset Ha$.

Since $Ha \subset [a]$ and $[a] \subset Ha$, $Ha = [a] = \{x \in G \mid a \cong x \mod H\}$. $\qquad\square$

Theorem 2.1 (Lagrange's theorem) *For any finite group G, the order of every subgroup H of G divides the order of G.*

Proof Considering that $[a] = Ha$, the equivalence classes to form a partition of G, i.e.,

$$[a_1] \cup [a_2] \cup \ldots \cup [a_k] = G$$

where $[a_i] \cap [a_j] \neq \emptyset$. The right lateral classes form a partition given by

$$G = Ha_1 \cup Ha_2 \cup \ldots \cup Ha_i \cup \ldots \cup Ha_k, \qquad 1 \leq i \leq k$$

Now, we establish a bijective map:

$$Ha_i \to H$$
$$ha_i \longmapsto H$$

that is to say $\mid Ha_i \mid = \mid H \mid \qquad \forall\, 1 \leq i \leq k$. Then

$$\mid G \mid = \mid Ha_1 \mid + \mid Ha_2 \mid + \ldots + \mid Ha_k \mid = \mid H \mid + \mid H \mid + \ldots + \mid H \mid = k \mid H \mid, k \in \mathbb{Z}$$

This means that the order of every subgroup H of G divides the order of G, i.e.,

$$|H|/|G|$$ □

Definition 2.9 If G is a finite group and H is a subgroup of G, the index of H in G is

$$i_G(H) = \frac{|G|}{|H|}$$

Definition 2.10 Let G a finite group and $a \in G$. The order of a is the minimum positive integer n such that $a^n = e$. The order of a is denoted by $O(a)$, then $a^{O(a)} = e$ [14, 16].

Proposition 2.4 *Let G a finite group and $a \in G$, then*

$$O(a)/|G|$$

Proof Let suppose $H = \langle a \rangle$, then $O(a) = H$. By Lagrange's theorem, $|H|/|G|$, hence $O(a)/|G|$. □

Corollary 2.1 *If G is a finite group of order n, then $a^n = e$ $\forall a \in G$.*

Proof According to above proposition, $O(a)/|G|$. By hypothesis, the order of G is n, then $O(a)/n$, so there exists $k \in \mathbb{Z}$ such that $n = kO(a)$. Note that $a^n = a^{kO(a)} = \left[a^{O(a)}\right]^k = e^k = e$. □

Definition 2.11 A group N of G is said to be a normal subgroup of G, denoted by $N \lhd G$, if for all g in G and for all $n \in N$, $gng^{-1} \in N$.

Lemma 2.2 *N is a normal subgroup of G if and only if $gNg^{-1} = N$ for all g in G.*

Proof 1. (Necessity). If $gNg^{-1} = N$ for all g in G, in particular, $gNg^{-1} \subset N$, then $gng^{-1} \in N$ for all $n \in N$. Therefore N is a normal subgroup of G.
2. (Sufficiency). If N is a normal subgroup of G, then $gng^{-1} \in N$. Let $g \in G$, then for all n in N, $gNg^{-1} \subset N$. On the other hand, $g^{-1}Ng = g^{-1}N(g^{-1})^{-1} \subset N$, besides $N = eNe = \left(gg^{-1}\right)N\left(gg^{-1}\right) = g\left(g^{-1}Ng\right)g^{-1} \subset gNg^{-1}$. Therefore $gNg^{-1} = N$. □

Lemma 2.3 *Let N a subgroup of G. N is a normal subgroup of G if and only if all left lateral class of N in G is a right lateral class of N in G.*

Proof Let $aH = \{ah \mid h \in H\}$ a left lateral class.

1. (Sufficiency). If N is a normal subgroup of G for all $g \in G, n \in N$, then from the above result, $gNg^{-1} = N$. Further $gN = gNe = gN\left(g^{-1}g\right) = \left(gNg^{-1}\right)g = Ng$. Therefore, all left lateral class coincides with the right lateral class.
2. (Necessity). Now, let suppose $Ng = gN$, then $gNg^{-1} = (gN)g^{-1} = (Ng)g^{-1} = N\left(gg^{-1}\right) = Ne = N$.

Hence that $N \lhd G$.   □

Definition 2.12 Let H a subgroup of G. Let

$$HH = \{h_1 h_2 \mid h_1, h_2 \in H\} \subset H$$

The proof of this assertion is immediate.

Remark 2.3 $HH = H$. It is clear that $HH \subset H$. On the other hand, chose $h_1 h_2 = h \in H$ since $H < G$ as well as $H \subset HH$. Therefore $HH = H$.

Corollary 2.2 *Let N a normal subgroup of G, $a, b \in G$, then*

$$NaNb = N(Na)b = NNab = Nab, \quad ab \in G$$

i.e., the product of right lateral classes is a right lateral class.   □

Definition 2.13 Let G/N the collection of right lateral classes of N in G:

$$G/N = \{Na \mid a \in G\}$$

Theorem 2.2 *Let G a group, N is a normal subgroup of G, then G/N is a group called* **quotient group.**

Proof Let $x, y, z \in G/N$, then $x = Na$, $y = Nb$ and $z = Nc$ for $a, b, c \in G$.

1. (Closure) Let $a, b \in G \Rightarrow Nab \in G/N$.
2. (Associativity) Note that $(xy)z = (x * y) * z = (NaNb)Nc = (Nab)Nc = Nabc = Na(Nbc) = Na(NbNc) = x * (y * z) = x(yz)$.
3. (Existence of identity element). Let $N = Ne$ then

$$xN = NaNe = NN(ae) = NNa = Na = x$$
$$Nx = NeNa = NN(ea) = NNa = Na = x$$

4. (Existence of inverse element) Let $Na^{-1} \in G/N$. To show that $x^{-1} = Na^{-1}$ is the inverse element of $x = Na$. Indeed, as $x \in G/N$ then $x = Na$ for any a in G. Then

$$xx^{-1} = NaNa^{-1} = NNaa^{-1} = NNe = Ne = N$$
$$x^{-1}x = Na^{-1}Na = NNa^{-1}a = NNe = Ne = N$$

hence $x^{-1} = Na^{-1}$ and G/N is a group.   □

2.3 Homomorphisms

Definition 2.14 Let $\varphi : G \to \bar{G}$. Let $(G, \circ)$ and $(\bar{G}, \Box)$ groups with $\circ$ and $\Box$ operations defined G and $\bar{G}$ respectively. It is said that φ is a homomorphism if for all $a, b \in G$:

$$\varphi(a \circ b) = \varphi(a) \Box \varphi(b) \tag{2.1}$$

Example 2.3 Let $G = \mathbb{R}^+$ with usual multiplication operation in $\mathbb{R}$ and $\bar{G} = \mathbb{R}^+$ with usual addition operation in $\mathbb{R}$. Let $\varphi : \mathbb{R}^+ \to \mathbb{R}^+$ with $\varphi(r) = \ln r$. Let $r_1, r_2 \in \mathbb{R}^+$, then:

$$\varphi(r_1 \cdot r_2) = \ln(r_1 \cdot r_2) = \ln r_1 + \ln r_2 = \varphi(r_1) + \varphi(r_2)$$

thus φ is a homomorphism.

Lemma 2.4 *Let G a group and N a normal subgroup of G. Let $\varphi : G \to G/N$ such that $\varphi(x) = Nx$ then φ is a homomorphism.*

Proof Let $x, y \in G$ then $\varphi(x) = Nx$ and $\varphi(y) = Ny$. Note that

$$\varphi(xy) = Nxy = NNxy = NxNy = \varphi(x)\varphi(y)$$

Hence φ is a homomorphism. $\Box$

Definition 2.15 If φ is a homomorphism of G in $\bar{G}$. The kernel of φ, denoted by K_φ (or Ker φ) is given by:

$$K_\varphi = \{x \in G \mid \varphi(x) = \bar{e}\} \subset G \tag{2.2}$$

where $\bar{e}$ is the identity of $\bar{G}$.

Lemma 2.5 *Let φ a homomorphism of G in $\bar{G}$, then:*

1. $\varphi(e) = \bar{e}$
2. $\varphi(x^{-1}) = [\varphi(x)]^{-1} \; \forall \, x \in G$

Proof 1. Let $x \in G$. By the identical element axiom, $x = xe$ then $\varphi(x) = \varphi(xe) = \varphi(x)\varphi(e)$. On the other hand $\varphi(x) = \varphi(x)\bar{e}$. Finally by the left cancellation rule we have $\varphi(e) = \bar{e}$.
2. Note that $\bar{e} = \varphi(e) = \varphi(xx^{-1}) = \varphi(x)\varphi(x^{-1})$ and $\bar{e} = \varphi(e) = \varphi(x^{-1}x) = \varphi(x^{-1})\varphi(x)$ therefore $\varphi(x^{-1})$ is the inverse element of $\varphi(x)$, this is $\varphi(x^{-1}) = [\varphi(x)]^{-1} \; \forall \, x \in G$. $\Box$

Example 2.4 1. Let G an Abelian group and $\varphi : G \to G$ such that $\varphi(a) = a^2$. Let $a_1, a_2 \in G$ such that $\varphi(a_1) = a_1^2$ and $\varphi(a_2) = a_2^2$. Furthermore as G is Abelian, $a_1 a_2 = a_2 a_1$ then it follows that $(a_1 a_1)(a_2 a_2) = (a_1 a_2)(a_1 a_2)$. Finally $\varphi(a_1 a_2) = (a_1 a_2)^2 = (a_1 a_2)(a_1 a_2) = (a_1 a_1)(a_2 a_2) = a_1^2 a_2^2 = \varphi(a_1)\varphi(a_2)$. Therefore φ is a homomorphism.

2. Let $(G, *)$, $(G', \triangle)$ groups and e' the identity element of G'. The function $\varphi : G \to G'$ defined by $\varphi(g) = e'$ for all $g \in G$ is a homomorphism. To verify this statement, let $g_1, g_2 \in G$. It is not hard to see that $\varphi(g_1 * g_2) = e' = e' \triangle e' = \varphi(g_1) \triangle \varphi(g_2)$.

3. Let $(G, *)$ a group. The identity function $\text{Id}_G : G \to G$ is a homomorphism. Indeed, let $g_1, g_2 \in G$ then $\text{Id}_G(g_1) = g_1$ and $\text{Id}_G(g_2) = g_2$. Now, let $g_1 * g_2 \in G$ then $\text{Id}_G(g_1 * g_2) = g_1 * g_2 = \text{Id}_G(g_1) * \text{Id}_G(g_2)$. This prove that φ is a homomorphism.

4. Let $(\mathbb{Z}, +)$ and $(G', \cdot)$ groups, where $G' = \{1, -1\}$ and $\cdot$ is the usual product between integers. Let define $\varphi : \mathbb{Z} \to G'$ in the following form:

$$\varphi(n) = \begin{cases} 1, & \text{if } n \text{ is even} \\ -1, & \text{if } n \text{ is odd} \end{cases}$$

Is $\varphi(n)$ a homomorphism? To answer this question, it is necessary to consider three cases:

a. If x_1, x_2 are even numbers then these numbers can be represented as $x_1 = 2k_1$ and $x_2 = 2k_2$, where $k_1, k_2 \in \mathbb{Z}$. According to $\varphi(n)$, $\varphi(x_1) = 1$, $\varphi(x_2) = 1$. Further, $x_1 + x_2 = 2k_1 + 2k_2 = 2(k_1 + k_2)$ is even then $\varphi(x_1 + x_2) = 1 = 1 \cdot 1 = \varphi(x_1) \cdot \varphi(x_2)$.

b. If x_1, x_2 are odd numbers then they can be represented in the form $x_1 = 2k_1 + 1$ and $x_2 = 2k_2 + 1$, where $k_1, k_2 \in \mathbb{Z}$. Further, $x_1 + x_2 = (2k_1 + 1) + (2k_2 + 1) = 2(k_1 + k_2 + 1)$ is even, $\varphi(x_1) = -1$ and $\varphi(x_2) = -1$, then $\varphi(x_1 + x_2) = 1 = (-1) \cdot (-1) = \varphi(x_1) \cdot \varphi(x_2)$.

c. If x_1 is even and x_2 is a odd number, the sum $x_1 + x_2 = 2k_1 + 2k_2 + 1 = 2(k_1 + k_2) + 1$ is a odd number, $\varphi(x_1) = 1$, $\varphi(x_2) = -1$. On the other hand $\varphi(x_1 + x_2) = -1 = (1) \cdot (-1) = \varphi(x_1) \cdot \varphi(x_2)$.

Hence $\varphi(n)$ is a homomorphism.

5. Let $n \in \mathbb{N}$, $G = \mathbb{Z}$ and $G' = \mathbb{Z}/n\mathbb{Z} = \mathbb{Z}_n$. Let define $\varphi : \mathbb{Z} \to \mathbb{Z}_n$ by $\varphi(a) = [a]$. This function is a homomorphism because $\varphi(a + b) = [a + b] = [a] + [b] = \varphi(a) + \varphi(b)$.

6. Let $\mathbb{C}^* = \mathbb{C} \setminus \{0\}$ and the groups $(\mathbb{C}^*, *)$ and $(\mathbb{R}^+, \cdot)$. Operations $*$ and $\cdot$ represent the usual product of complex numbers and real numbers respectively. Let $\varphi : \mathbb{C}^* \to \mathbb{R}^+$ the function given by $\varphi(z) = |z|$. Show that φ is a homomorphism. The proof is left to the reader.

Definition 2.16 A homomorphism $\varphi : G \to G'$ is said that is

- Monomorphism if it is an injective function (one to one or 1–1).
- Epimorphism if it is surjective (onto).
- Isomorphism if it is a bijection (injective and surjective).

Definition 2.17 An automorphism of a group one means an isomorphism of a group with itself.

Definition 2.18 If $\varphi : G \to G'$ is a isomorphism then G and G' are isomorphic. This characteristic is denoted by $G \cong G'$.

Proposition 2.5 *If $\varphi : G \to G'$ is a homomorphism then Im $\varphi < G'$, where Im $\varphi = \{y \in G' \mid \varphi(x) = y,\ x \in G\} \subset G'$.*

Proof Let $y_1, y_2 \in$ Im φ then $y_1 = \varphi(x_1)$, $y_2 = \varphi(x_2) \in G'$, $x_1, x_2 \in G$. In addition $y_1 y_2 = \varphi(x_1)\varphi(x_2) = \varphi(x_1 x_2)$ because φ is a homomorphism, then $y_1 y_2 \in$ Im φ. Now, let $y = \varphi(x) \in$ Im φ. By hypotheses φ is a homomorphism then $\varphi(x)\varphi(x^{-1}) = \varphi(xx^{-1}) = \varphi(e) = e' \in G'$. In the same manner $\varphi(x^{-1})\varphi(x) = \varphi(x^{-1}x) = \varphi(e) = e' \in G'$. Therefore $[\varphi(x)]^{-1} = \varphi(x^{-1})$ and Im $\varphi < G'$. This proves the assertion. $\qquad\square$

Proposition 2.6 *Let $\varphi : G \to G'$ a homomorphism then φ is a monomorphism if and only if Ker $\varphi = \{0\}$.*

Proof Let x_1, x_2 such that $\varphi(x_1) = \varphi(x_2)$ then $\varphi(x_1) - \varphi(x_2) = 0$. From last equality we have $\varphi(x_1 - x_2) = 0$ because φ is a homomorphism. According to the hypothesis Ker $\varphi = \{0\}$ then $x_1 - x_2 \in$ Ker $\varphi = \{0\}$. Therefore $x_1 - x_2 = 0$ and this implies that $x_1 = x_2$. This proves the sufficiency of the proposition. The necessity is left to the reader. $\qquad\square$

Theorem 2.3 *If $\varphi : G \to G'$ a homomorphism then*

1. *Ker $\varphi < G$*
2. *For all $a \in G$, a^{-1} (Ker φ) $a \subset$ Ker φ*

Proof Ker $\varphi \neq \emptyset$ since exists $e \in G$ such that $\varphi(e) = e'$. To prove the first assertion, let $x, y \in$ Ker φ then $\varphi(xy) = \varphi(x)\varphi(y) = e'e' = e'$. This implies that $xy \in$ Ker φ and therefore Ker φ is closed. On the other hand, let $x \in$ Ker φ then $\varphi(x) = e'$. Note that $\varphi(x^{-1}) = e'\varphi(x^{-1}) = \varphi(x)\varphi(x^{-1}) = \varphi(xx^{-1}) = \varphi(e) = e'$. Therefore $x^{-1} \in$ Ker φ and the first assertion is proved. For the second one, let $a \in G$, $g \in$ Ker φ then $\varphi(a^{-1}ga) = \varphi(a^{-1})\varphi(g)\varphi(a) = \varphi(a^{-1})e'\varphi(a) = \varphi(a^{-1})\varphi(a) = \varphi(a^{-1}a) = \varphi(e) = e'$. This implies that $a^{-1}ga \in$ Ker φ for all $g \in G$. Therefore a^{-1} (Ker φ) $a \subset$ Ker φ. In addition, Ker $\varphi \lhd G$. $\qquad\square$

Example 2.5 Let $\varphi : \mathbb{Z} \to \mathbb{Z}$ given by $\varphi(a) = 2a$. Is φ a homomorphism? If φ is homomorphism, is it a monomorphism? To answer the first question, let $a, b \in \mathbb{Z}$ then $\varphi(a) = 2a$, $\varphi(b) = 2b$. On the other hand, $\varphi(a + b) = 2(a + b) = 2a + 2b = \varphi(a) + \varphi(b)$, then φ is a homomorphism under the sum in $\mathbb{Z}$. Is left as an exercise to verify that φ is not homomorphism under the product in $\mathbb{Z}$. Note that Ker $\varphi = \{a \in \mathbb{Z} \mid \varphi(a) = 2a = 0\}$ then Ker $\varphi = \{0\}$. Thus φ is a monomorphism.

Exercise 2.4 Check if the following functions are homomorphism with the sum and product. In addition to check if they are also monomorphisms.

1. $\varphi : \mathbb{Z} \to G'$ where $G' = \{1, -1\}$ and

$$\varphi(n) = \begin{cases} 1, & \text{if } n \text{ is even} \\ -1, & \text{if } n \text{ is odd} \end{cases}$$

2. $\varphi : \mathbb{C}^* \to \mathbb{C}^*$ the function given by $\varphi(z) = |z|$ where $\mathbb{C}^* = \mathbb{C} \setminus \{0\}$.

Proposition 2.7 *Let G a group and N a normal subgroup of G. There exist a epimorphism $\varphi : G \to G/N$ with Ker $\varphi = N$.*

Proof Let define $\varphi : G \to G/N$ given by $\varphi(a) = [a]$. Let $a, b \in G$ then $\varphi(ab) = [ab] = [a][b] = \varphi(a)\varphi(b)$ so that φ is a homomorphism. On the other hand φ is onto (by construction) then φ is a epimorphism because if $a \in$ Ker φ then $\varphi(a) = e$ then $a \cong e \mod N$. This implies that $ae^{-1} \in N$ hence $a \in N$. The inverse demonstration is followed immediately and therefore Ker $\varphi = N$. $\qquad\square$

2.4 The Isomorphism Theorems

Theorem 2.4 (First theorem of isomorphisms) *Let $\varphi : G \to G'$, $g \in G$, $\varphi(g) \in G'$ an epimorphism with kernel K then $G/K \cong G'$.*

Proof Let the function $\bar{\varphi} : G/K \to G'$ with $\varphi(\bar{k}g) = \varphi(g)$. By construction $\bar{\varphi}$ is onto then for all $\varphi(g) \in G'$ there exists $g \in G$ such that $\bar{\varphi}(kg) = \varphi(g)$ where $kg \in G/K$. To prove the injectivity of φ, we consider $g_1, g_2 \in G$ such that $\bar{\varphi}(kg_1) = \bar{\varphi}(kg_2)$ then we will verify that $kg_1 = kg_2$. According to $\bar{\varphi}$ function, if $\bar{\varphi}(kg_1) = \bar{\varphi}(kg_2)$ then $\varphi(g_1) = \varphi(g_2)$ so that $\varphi(g_1)\varphi(g_2^{-1}) = e'$ and therefore $\varphi(g_1g_2^{-1}) = e'$. Hence $g_1g_2^{-1} \in K$ and $g_2g_1^{-1} \in K$. From these assertions it follows that $g_1 \cong g_2 \mod K$ and $g_2 \cong g_1 \mod K$. Therefore $g_1 \sim g_2$, $g_2 \sim g_1$ and $[g_1] = [g_2]$ then $kg_1 = kg_2$, this yields to $\bar{\varphi}$ is injective. The last part of the proof is to check that $\bar{\varphi}$ is a homomorphism. To check this, note that $\bar{\varphi}(kg_1kg_2) = \bar{\varphi}(kkg_1g_2) = \bar{\varphi}(kg_1g_2) = \varphi(g_1g_2) = \varphi(g_1)\varphi(g_2) = \bar{\varphi}(kg_1)\bar{\varphi}(kg_2)$. Hence $\bar{\varphi}$ is a isomorphism and $G/K \cong G'$. $\qquad\square$

Theorem 2.5 (Second theorem of isomorphisms) *Let G a group, $H < G$ and $N \lhd G$ then $HN < G$, $H \cap N \lhd H$, $N \lhd HN$ and $HN/N \cong H/(H \cap N)$.*

Theorem 2.6 (Third theorem of isomorphisms) *Let G a group, $N \lhd G$ and $K < N$ with $K \lhd G$ then $G/K/N/K \cong G/N$.*

Definition 2.19 Let $G_1, G_2, \ldots, G_n$ groups. The **direct product** (or exterior product) denoted by $G_1 \times G_2 \times \cdots \times G_n$ is the set $(a_1, a_2, \ldots, a_n)$ where each $a_i \in G$ for all $i \in \mathbb{N}$ and the operation in the direct product is defined component by component:

$$(a_1, a_2, \ldots, a_n)(b_1, b_2, \ldots, b_n) = (a_1b_1, a_2b_2, \ldots, a_nb_n) \tag{2.3}$$

Remark 2.4

$$G = G_1 \times G_2 \times \cdots \times G_n$$

is a group whose identity element is $(e_1, e_2, \ldots, e_n)$. The inverse element of $(a_1, a_2, \ldots, a_n)$ is given by $(a_1^{-1}, a_2^{-1}, \ldots, a_n^{-1})$.

References

1. Birkhoff, G., Mac Lane, S.: A Survey of Modern Algebra. Universities Press (1965)
2. Dummit, D.S., Foote, R.M.: Abstract Algebra (Vol. 3). Hoboken: Wiley (2004)
3. Fraleigh, J.B.: A First Course in Abstract Algebra. Addison-Wesley (2003)
4. Hungerford, T.W.: Abstract Algebra: an Introduction. Cengage Learning (2012)
5. Sontag, E.D.: Linear systems over commutative rings. A Survey. Ricerche di Automatica 7(1), 1–34 (1976)
6. Bourbaki, N.: Theory of Sets. Springer, Berlin, Heidelberg (2004)
7. Herstein, I.N.: Topics in Algebra. Wiley (2006)
8. Lang, S.: Algebra. Springer (2002)
9. Herstein, I.N.: Abstract Algebra. Prentice Hall (1996)
10. Saracino, D.: Abstract Algebra: a First Course. Waveland Press (2008)
11. Jacobson, N.: Lectures in Abstract Algebra, vol. 1. Van Nostrand, Basic Concepts (1951)
12. Zariski, O., Samuel, P.: Commutative Algebra Vol. 1. Springer Science & Business Media (1958)
13. Zariski, O., Samuel, P.: Commutative Algebra Vol. 2. Springer Science & Business Media (1960)
14. Kunz, E.: Introduction to Commutative Algebra and Algebraic Geometry. Birkhuser (1985)
15. Borel, A.: Linear Algebraic Groups (Vol. 126). Springer Science & Business Media (2012)
16. Samuel, P., Serge, B.: Mthodes d'algbre abstraite en gomtrie algbrique. Springer (1967)

Chapter 3
Rings

Abstract This chapter contains introductory concepts to ring theory, beginning with basic definitions, followed by the definitions of ideals, homomorphisms and rings, then isomorphism theorems in rings are given, next properties about integer rings are introduced and finally polynomial rings are explained.

3.1 Basic Definitions

Definition 3.1 A nonempty set R together with two binary operators $+$ and $*$ is called **ring** if satisfies the following axioms [1–5]:

1. The operation $a + b \in R$ for all $a, b \in R$.
2. $a + (b + c) = (a + b) + c$ for all $a, b, c \in R$.
3. $a + b = b + a$ for all $a, b \in R$.
4. There exists $0 \in R$ such that $a + 0 = a$ for all $a \in R$.
5. For all $a \in R$ there exists $b \in R$ such that $a + b = 0$. In this case $b = -a$.
6. Given $a, b \in R$, $a * b \in R$ for all $a, b \in R$.
7. $a * (b * c) = (a * b) * c$ for all $a, b, c \in R$.
8. $a * (b + c) = a * b + a * c$ and $(b + c) * a = b * a + c * a$ for all $a, b, c \in R$.

Remark 3.1 The axioms 1–5 listed above correspond to an additive Abelian group under addition. Axioms 6 and 7 mean that a ring is a semigroup (closed and associative). Finally the last two axioms correspond to the distributivity property.

Definition 3.2 It is said that a ring R is a **ring with identity** if there exists $1 \in R$, $1 \neq 0$ such that $a * 1 = 1 * a = a$ for all $a \in R$.

Definition 3.3 A ring R is a **commutative ring** if $a * b = b * a$ for all $a, b \in R$.

© Springer Nature Switzerland AG 2019

R. Martínez-Guerra et al., *Algebraic and Differential Methods for Nonlinear Control Theory*, Mathematical and Analytical Techniques with Applications to Engineering, https://doi.org/10.1007/978-3-030-12025-2_3

Definition 3.4 Let R a ring, $a \in R$, $a \neq 0$. It said that a **is a divisor of zero** if there exists $b \in R$, $b \neq 0$ such that $a * b = 0$ (right division) or there exists $c \in R$, $c \neq 0$ such that $c * a = 0$ (left division).

Example 3.1 In $\mathbb{Z}_6$ we have $[3][2] = [0]$ then $[3]$ and $[2]$ are divisors of zero.

Definition 3.5 Let R a ring with identity. It said that R is a **ring with division** if for all $a \in R$, $a \neq 0$ there exists $b \in R$ such that $ab = ba = 1$. In this case $b = a^{-1}$.

Example 3.2 In $\mathbb{Z}_5$, $[2][3] = 1$ then $[2] = [3]^{-1}$. Analogously $[3][2] = [1]$ then $[3] = [2]^{-1}$.

Definition 3.6 A **field** is a ring with division and it is also commutative.

Remark 3.2 A field is an Abelian group with the sum and the multiplication.

Example 3.3 $\mathbb{Z}$ is not a field under the multiplication and sum. The proof is left to the reader as an exercise.

Definition 3.7 A commutative ring with identity is an **integral domain** if $ab = 0$ then $a = 0$ or $b = 0$. In other words, an integral domain is a nonzero commutative ring with no nonzero zero divisors.

Example 3.4 1. In $\mathbb{Z}_6$, $[2][3] = [6] = [0]$. Note that $[2] \neq [0]$ and $[3] \neq [0]$ therefore is not an integral domain.
   2. $\mathbb{Z}_5$ is a commutative ring with identity and every element different from $[0]$ has inverse, for example $[1]^{-1} = [1]$ or $[2]^{-1} = [3]$. In addition, $\mathbb{Z}_5$ is an integral domain. Is it a field?

Remark 3.3 If p is prime number then $\mathbb{Z}_p$ is a field.

Exercise 3.1 Let

$$M = M_{2\times 2}(\mathbb{R}) = \left\{ \begin{bmatrix} a & b \\ c & d \end{bmatrix} \middle| a, b, c, d \in \mathbb{R} \right\}$$

with usual sum and multiplication of matrices in $\mathbb{R}$. Let $(I_2)_{ij} = \delta_{ij}$ the identity matrix, where

$$\delta_{ij} = \begin{cases} 1, & \text{if } i = j \\ 0, & \text{otherwise.} \end{cases}$$

Verify that M is a ring with identity, not commutative and that is not an integral domain.

Lemma 3.1 (Cancellation rules) *Let $a, b, c \in R$ if $b + a = c + a$ then $b = c$.*

Proof Applying directly the axioms of a ring:

$$b = b + 0 = b + (a + (-a)) = (b + a) + (-a) = (c + a) + (-a)$$
$$= c + (a + (-a)) = c + 0 = c$$

therefore $b = c$. $\qquad\square$

Proposition 3.1 *Let R be a ring, $a, b \in R$ then [6, 7]:*

1. $a \cdot 0 = 0 \cdot a = 0$
2. $a(-b) = (-a)(b) = -(ab)$
3. $(-a)(-b) = (-(-a))b = ab$
4. *If* $1 \in R$ *then* $(-1)a = -a$

Proof Applying the axioms of a ring, mainly considering the additive inverse we have:

1. $a \cdot 0 + 0 = a \cdot 0 = a \cdot (0 + 0) = a \cdot 0 + a \cdot 0$ then by cancellation rule $a \cdot 0 = 0$.
2. $a(-b) + a \cdot b = a(-b + b) = a \cdot 0 = 0$, then $a(-b) = -ab$.
3. $(-a)(-b) + (-a)b = (-a)(-b + b) = -a \cdot 0 = 0$, then $(-a)(-b) = -(-a)b$.
4. $(-1)a + a = a(-1 + 1) = a \cdot 0 = 0$ then $(-1)a = -a$. $\qquad\square$

3.2 Ideals, Homomorphisms and Rings

Definition 3.8 A function $\varphi : R \to R'$ is a homomorphism if [1, 3–5, 8]

1. $\varphi(a) + \varphi(b) = \varphi(a + b)$
2. $\varphi(a)\varphi(b) = \varphi(a \cdot b)$

Definition 3.9 Let $\varphi : R \to R'$ a homomorphism of rings [9, 10], then

1. φ is monomorphism if it is injective (one-to-one or $1 - 1$).
2. φ is epimorphism if it is surjective (onto).
3. φ is isomorphism if it is bijection ($1 - 1$ and onto).

Definition 3.10 The kernel of φ is Ker $\varphi = \{x \in R \mid \varphi(x) = 0\} \subset R$.

Proposition 3.2 *Let $\varphi : R \to R'$ a homomorphism of rings, then [8, 11]*

1. Ker φ *is an additive subgroup.*
2. *If $k \in$ Ker φ and $r \in R$ then $rk, kr \in$ Ker φ.*
3. φ *is $1 - 1$ if and only if Ker $\varphi = \{0\}$.*

Proof We only prove the second assertion. Let $k \in$ Ker φ and $r \in R$ then $\varphi(rk) = \varphi(r) \cdot \varphi(k) = \varphi(r) \cdot 0 = 0$. In the same manner $\varphi(kr) = \varphi(k) \cdot \varphi(r) = 0 \cdot \varphi(r) = 0$, thus $rk, kr \in$ Ker φ. The proof of 1 and 3 are left to the reader. $\qquad\square$

Definition 3.11 Let R a ring. A subset I of R is said to be an **ideal** de R if

1. I is an additive subgroup of R.
2. Let $r \in R$, $a \in I$ then $ra \in I$ and $ar \in I$ (absorption property).

Corollary 3.1 *If $\varphi : R \to R'$ is an homomorphism, then Ker φ is an ideal.*

Proof It is immediate from **Proposition** 3.2.   □

Definition 3.12 Let R a ring [12–14], I an ideal of R then R/I (**quotient ring**) is a group with the sum of equivalence classes:

$$(a + I) + (b + I) = (a + b) + I, \ a, b \in R$$

and we define the product as:

$$(a + I)(b + I) = ab + I$$

Remark 3.4 Let R a ring then $\{0\}$ and R are rings of R called the trivial ideals.

Definition 3.13 If I is an ideal of R and $I \neq R$, it is said that I is a **proper ideal**.

Remark 3.5 Let R a ring with identity 1, I an ideal of R such that $1 \in I$, then $I = R$. Since I is an ideal of R then $I \subset R$ we should verify that $R \subset I$. Let $a \in R$, since $a \cdot 1 = a$ by definition $a = a \cdot 1 \in I$ ($1 \in I$) then $R \subset I$. Therefore $R = I$.

Example 3.5 Let R be a commutative ring with identity. Let $a \in R$ and let define $(a) = \{ax \mid x \in R\}$. Prove that (a) is an ideal in R. It is left to the reader as an exercise.

Definition 3.14 Let R a commutative ring with identity. A **principal ideal** is an ideal of the form (a) for some $a \in R$.

Definition 3.15 Let R an integral domain. It is said that R is a **principal ideals domain** if every ideal of R is principal.

Exercise 3.2 Let I, J ideals of a ring R. Let define $I + J = \{a + b \mid a \in I, b \in J\}$. Show that $I + J$ and $I \cap J$ are ideals.

Remark 3.6 The quotient ring R/I is a ring with the addition and product respectively:

- $(a + I) + (b + I) := (a + b) + I$
- $(a + I)(b + I) := ab + I$

We have the following assumption:

Lemma 3.2 *The function*

$$\varphi : R \to R/I$$
$$a \mapsto a + I$$

is an epimorphism.

Proof Note that φ is a homomorphism. Indeed, $\varphi(a + b) = (a + b) + I = (a + I) + (b + I) = \varphi(a) + \varphi(b)$. On the other hand, $\varphi(ab) = ab + I = (a + I)(b + I) = \varphi(a)\varphi(b)$. By construction, function φ is onto. Thus φ is an epimorphism. $\square$

Exercise 3.3 Show that kernel of φ given by Ker $\varphi = \{a \in R \mid \varphi(a) = 0 + I\}$ coincides with the ideal I.

3.3 Isomorphism Theorems in Rings

Now we present three theorems about isomorphisms in ring theory.

Theorem 3.1 (First theorem of isomorphisms) *Let* $\varphi : R \to R'$ *an epimorphism of rings and* $K = Ker\, \varphi$ *then* R/K *is isomorphic to* R', *i.e.,* $R/K \cong R'$.

Theorem 3.2 (Second theorem of isomorphisms) *Let* R *a ring,* A *subset of* R (A *is a sub-ring)and* B *and ideal of* R *then* $A + B$ *is a sub-ring of* R *and* $A \cap B$ *is an ideal of* A. *In addition*

$$(A + B)/B \cong A/(A \cap B)$$

Theorem 3.3 (Third theorem of isomorphisms) *Let* I, J *ideals of ring* R *with* $I \subset J$ *then* J/I *is an ideal of* R/I. *In addition*

$$(R/I)/(J/I) \cong R/J$$

Remark 3.7 Let $\mathbb{F}$ a field (commutative ring with division), $\{0\}$ is an ideal of F.

Example 3.6 Let I an ideal of F, $I \neq \{0\}$ then $F = I$.

Proof Let $a \in I$, $a \neq 0$ then $1 = a^{-1}a \in I$. On the other hand, let $r \in F$ then $r = 1 \cdot r \in I$ hence $F \subset I$ thus $F = I$.

Example 3.7 Let $n \in \mathbb{N}$, $n > 1$ and

$$\varphi : \mathbb{Z} \to \mathbb{Z}/n\mathbb{Z}$$
$$a \mapsto [a]$$

φ is an epimorphism. Since Ker $\varphi = \{a \in \mathbb{Z} \mid [a] = [0]\} = \{a \in \mathbb{Z} \mid a \cong 0 \mod n\} = \{a \in \mathbb{Z} \mid a = nz, z \in \mathbb{Z}\} = n\mathbb{Z}$.

Example 3.8 Let $R = \{f : [0, 1] \to \mathbb{R} \mid f \text{ is continuous}\}$. We define:

$$(f + g)(x) = f(x) + g(x)$$
$$(f \cdot g)(x) = f(x) \cdot g(x)$$

R is a commutative ring with identity. The proof is left to the reader.

Example 3.9 Let R a group, $I = \{f \in R \mid f(\frac{1}{2}) = 0\}$. I is an ideal. Show that $R/I \cong R$. Define

$$\varphi : R \to R$$
$$f \mapsto f(\tfrac{1}{2})$$

φ is an homomorphism. Let $f, g \in R$ then

$$\varphi(f + g) = (f + g)(\tfrac{1}{2}) = f(\tfrac{1}{2}) + g(\tfrac{1}{2}) = \varphi(f) + \varphi(g)$$
$$\varphi(f \cdot g) = (f \cdot g)(\tfrac{1}{2}) = f(\tfrac{1}{2}) \cdot g(\tfrac{1}{2}) = \varphi(f) \cdot \varphi(g)$$

it is clear that φ is an epimorphism (by construction). Hence by using Theorem 3.1 $R/I \cong R$.

Remark 3.8 In Chap. 10, we will focus on algebraic structures with a derivation. In the case of rings, any commutative ring can be viewed as a differential ring.

3.4 Some Properties of Integers

3.4.1 Divisibility

Definition 3.16 Let $a, b \in \mathbb{Z}$. It is said that b divides a if there exists $q \in \mathbb{Z}$ such that $a = bq$. In other words, b is a factor of a or a it is a multiple of b.

Remark 3.9 The notation for divisibility of numbers is the following:

- If b divides a, we write $b \mid a$.
- In other case, i.e., if b not divides a, we write $b \nmid a$.

Theorem 3.4 *Let $a, b, c \in \mathbb{Z}$.*

1. $b \mid b \,\forall\, b \in \mathbb{Z}$
2. $b \mid 0 \,\forall\, b \in \mathbb{Z} \setminus \{0\}$
3. $1 \mid a$ *and* $-1 \mid a \,\forall\, a \in \mathbb{Z}$
4. $0 \mid a$ *if and only if* $a = 0$
5. *If* $b \mid 1$ *then* $b = \pm 1$
6. *If* $b \mid a$ *and* $a \mid b$ *then* $a = \pm b$
7. *If* $b \mid a$ *and* $a \mid c$ *then* $b \mid c$
8. *If* $b \mid a$ *and* $a \mid c$ *then* $b \mid (a + c)$ *and* $b \mid (a - c)$
9. $b \mid a$ *then* $b \mid (ac) \,\forall\, c \in \mathbb{Z}$
10. $b \mid a$ *and* $b \mid c$ *then* $b \mid (as + ct) \,\forall\, s, t \in \mathbb{Z}$

11. $b \mid a \Longleftrightarrow b \mid -a \Longleftrightarrow -b \mid a \Longleftrightarrow -b \mid -a$
12. $b \mid a \Longleftrightarrow b \mid |a| \Longleftrightarrow |b| \mid a \Longleftrightarrow |b| \mid |a|$

Proof 1. According to definition of divisibility, we have to find a number $q \in \mathbb{Z}$ such that $b = b \cdot q$. Since $b = b \cdot 1$, it is obvious that $q = 1$ in the definition.
2. Since $0 = b \cdot 0$, in the definition of divisibility $q = 0$ for this case.
3. The results are immediately: $a = 1 \cdot a$ and $a = (-1)(a)$.
4. If 0 divides a then there exists $q \in \mathbb{Z}$ such that $a = 0 \cdot q$. It follows that $a = 0$. The rest of the proof is left to the reader.
5. By hypothesis, $b \mid 1$ then there exists $q \in \mathbb{Z}$ such that $1 = b \cdot q$ then $b \neq 0, q \neq 0$ and $1 = |bq|$. Since $b, q \in \mathbb{Z}$ and both numbers are not zero, then $|b| \geq 1$ and $|q| \geq 1$. If $|b| > 1$ then $|q| < |b||q| = 1$. This is a contradiction, then $b = |1|$, so that $b = 1$ or $b = -1$.
6. By hypothesis, there exist $q_1, q_2 \in \mathbb{Z}$ such that $a = bq_1$ and $b = aq_2$. Combining these expressions $a = (aq_2)q_1$. If $a = 0$ it is clear that $b = 0$ so that we consider $a \neq 0$ then $1 = q_1 q_2$. According to definition of divisibility, it follows that $q_1 \mid 1$ and by 5 we have that $q_1 = \pm 1$. Therefore $a = b(1) = b$ and $a = b(-1) = -b$.
7. If $b \mid a$ and $a \mid c$ then there exist $q_1, q_2 \in \mathbb{Z}$ such that $a = bq_1$ and $c = aq_2$. Combining these expressions $c = (bq_1)q_2 = b(q_1 q_2)$. Let $k = q_1 q_2 \in \mathbb{Z}$ then it follows that $b \mid c$.
8. Similar as above, let $q_1, q_2 \in \mathbb{Z}$ such that $a = bq_1$ and $c = aq_2$ (by hypothesis) then $a + c = bq_1 + bq_2 = b(q_1 + q_2)$. Since $q_1 + q_2 \in \mathbb{Z}$ then $b \mid (a + c)$. The rest of the proof is left to the reader.
9. By hypothesis there exists $q \in \mathbb{Z}$ such that $a = bq$. Let $c \in \mathbb{Z}$ then $ac = (bq)c = b(qc)$. Since $qc \in \mathbb{Z}$ it follows that $b \mid (ac)$.
10. If $b \mid a$ and $b \mid c$ then there exist $q_1, q_2 \in \mathbb{Z}$ such that $a = bq_1$ and $c = bq_2$. Let $s, t \in \mathbb{Z}$ then $as + ct = (bq_1)s + (bq_2)t = b(q_1 s + q_2 t)$. Since $q_1 s + q_2 t \in \mathbb{Z}$ we conclude that $b \mid (as + ct)$.

The proof of the statements 11 and 12 are left to the reader. □

Example 3.10 Let $n \in \mathbb{Z}$ such that $(-10) \mid n$ then $5 \mid n$. To prove this, according to the definition of divisibility, there exists $k \in \mathbb{Z}$ such that $n = (-10)k$. Since that $-10 = (5)(-2)$, then $n = 5(-2k)$ where $-2k \in \mathbb{Z}$. We have an integer number $m = -2k$ such that $n = 5m$. Therefore $5 \mid n$.

Example 3.11 Let $x \in \mathbb{Z}$ such that $(-12) \mid x$ and $x \neq 0$ then $|x| \geq 12$. To prove this, according to the definition of divisibility, there exists $m \in \mathbb{Z}$ such that $x = (-12)m$ with $m \neq 0$. If $m = 0$ then $x = 0$ that contradicts the hypothesis $x \neq 0$. Since $m \in \mathbb{Z}$ and $m \neq 0$, it follows that $|m| \in \mathbb{Z}$ and $|m| > 0$ so that $|m| \geq 1$. On the other hand $|x| = |-12m| = 12|m|$ and considering that $|m| \geq 1$ then $|x| = 12|m| \geq 12$. Therefore $|x| \geq 12$.

The above example can be generalized as follows.

Theorem 3.5 *Let $a, b \in \mathbb{Z}$ such that $b \neq 0$ and $a \mid b$ then $|a| \leq |b|$.*

Proof By definition, there exists $k \in \mathbb{Z}$ such that $b = ak$. If $k = 0$ then $b = 0$ that contradicts the hypothesis so that $k \neq 0$. Since $k \in \mathbb{Z}$ and $k \neq 0$ it follows that $|k| \in \mathbb{Z}$ and $|k| > 0$ then $|k| \geq 1$. Taking the absolute value of b and considering that $|k| \geq 1$ we have $|b| = |ak| = |a||k| \geq 1 \cdot |a| = |a|$. The proof is completed.   □

3.4.2 Division Algorithm

Theorem 3.6 *If $a, b \in \mathbb{Z}$ and $b \neq 0$, then there exist $q, r \in \mathbb{Z}$ (unique), such that:*

$$a = bq + r, \qquad 0 \leq r < |b|$$

Proof We define a set $S = \{a - bx \mid x \in \mathbb{Z}\}$. Firstly we will verify that $S \subset \mathbb{Z}^+ \cup \{0\}$.

(a) $b < 0$. If b is negative then $rb \leq -1$ this yields to $b|a| \leq -|a| \leq a$ then $a - b|a| \geq 0$.

(b) $b > 0$. If b is positive then $b \geq 1$. We have that $b(-|a|) \leq -|a| \leq a$ then $a - b(-|a|) \geq 0$.

Hence $S \subset \mathbb{Z}^+ \cup \{0\}$. On the other hand we state that $r < |b|$. Suppose that r is the smaller of the non-negative integers and $r \in S$, as well as we suppose that $r \geq |b|$ then $r - |b| \geq 0$, since $r \in S$, there exists $x = q \in \mathbb{Z}$ such that $r = a - bq$ then $r - |b| = a - bq - |b|$. We have the following

$$r - |b| = \begin{cases} a - b(q - 1), & \text{if } b > 0 \\ a - b(q + 1), & \text{if } b < 0 \end{cases}$$

but $r - |b| \in S$, so

- If $b < 0$ then $a - b(q - 1) = a - bq + b = r + b < r$.
- If $b > 0$ then $a - b(q + 1) = a - bq - b = r - b < r$.

Thus is not true that $r \geq |b|$ then $r < |b|$. The proof of uniqueness of q and r is left to the reader.   □

Remark 3.10 In the above algorithm, a is namely *dividend*, b is the *divisor* and the numbers q and r are *quotient* and *residue*, respectively.

3.4.3 Greatest Common Divisor

Definition 3.17 Let $a, b \in \mathbb{Z}$, not both zero. It is said that $d \in \mathbb{Z}$, $d > 0$, is the **greatest common divisor** (gcd) of a and b if:

1. $d \mid a$ and $d \mid b$
2. If $c \in \mathbb{Z}$ is such that $c \mid a$ and $c \mid b$, then $c \mid d$

Remark 3.11 If d is the greatest common divisor of a and b, it is very common to write $d = (a, b)$ or $d = gcd\{a, b\}$.

Lemma 3.3 *If $d = (a, b)$ then d is unique.*

Proof Suppose $d'(a, b)$, then from the Definition 3.17, $d \mid d'$ and $d' \mid d$. Finally, from 6 in Theorem 3.4, $d' = d$. □

Lemma 3.4 *Let $a, b \in \mathbb{Z}$, $b \neq 0$. If $b \mid a$, then*

$$|b| = (a, b)$$

Proof Consider statements 1. and 12. of **Theorem** 3.4. We have that $b \mid b$ and by hypothesis $b \mid a$ then $|b| \mid b$ and $|b| \mid a$ respectively. On the other hand, let $c \in \mathbb{Z}$ such that $c \mid a$ and $c \mid b$ then by the same Theorem, $c \mid |b|$. Therefore $|b| = (a, b)$. □

Lemma 3.5 *Let $a, b \in \mathbb{Z}$, not both zero. If $a = bq + r$ for some $q, r \in \mathbb{Z}$, then $d = (a, b)$ if and only if $d = (b, r)$.*

Proof If $b \nmid a$ then

$$a = bq + r, \ 0 \leq r \leq |b| \tag{3.1}$$

1. Let c such that $c \mid a$ and $c \mid b$ then there exist $\bar{q}, q' \in \mathbb{Z}$ such that $a = c\bar{q}, b = cq'$. From (3.1) $r = a - bq = c\bar{q} - cq'q = c(\bar{q} - q'q)$ then $c \mid r$.
2. Let l such that $l \mid b$ and $l \mid r$ then there exist $m, \bar{m} \in \mathbb{Z}$ such that $b = lm, r = l\bar{m}$. From (3.1) $r = a - bq = a - lmq = l\bar{m}$ then $a = l\bar{m} + lmq = l(\bar{m} + mq)$, i.e., $l \mid a$.

Thus $(a, b) = (b, r)$. □

Theorem 3.7 (The Euclidean Algorithm) *Let $a, b \in \mathbb{Z}$, not both zero, there exists $d = (a, b)$. In addition, d is the minimum positive integer such that, there exist $m, n \in \mathbb{Z}$ such that $d = am + bn$ (Bézout's identity).*

Proof The methodology to find the greatest common divisor is given by the following manner. If $b \nmid a$ then

$$a = bq + r, \ 0 \leq r < |b| \tag{3.2}$$

If $r \mid b$ then $b = qr$ and $r = (a, b)$. The algorithm is ended. On the other hand, if $r \nmid b$ then

$$b = q_1 r + r_1, \ 0 \leq r_1 < |r| \tag{3.3}$$

Suppose that $r_1 \mid r$ then $r_1 = (b, r)$ and the algorithm is finished. If $r_1 \nmid r$ then

$$r = r_1 q_2 + r_2, \ 0 \le r_2 < |r_1| \qquad (3.4)$$

Continuing the same process, in step $k + 1$ if $r_{k-1} \nmid r_{k-2}$ then $r_{k-2} = r_{k-1} q_k + r_k$, $0 \le r_k \le |r_{k-1}|$. Suppose that if $r_k \mid r_{k-1}$ then $r_{k-1} = r_k q_{k+1}$, therefore $r_k = (r_{k-2}, r_{k-1}) = \ldots = (r, r_1) = (b, r) = (a, b) = d$. On the other hand, from (3.2),

$$r = a - bq = a + (-q)b = m_1 a + n_1 b$$

where $m_1 = 1, n_1 = -q \in \mathbb{Z}$. From (3.3) making algebraic manipulations:

$$r_1 = b - r q_1 = b - (m_1 a + n_1 b) q_1 = -m_1 q_1 a + (1 - n_1 q_1)b = m_2 a + n_2 b,$$

where $m_2 = -m_1 q_1, n_2 = (1 - n_1 q_1) \in \mathbb{Z}$. In the same manner, from (3.4):

$$r_2 = m_3 a + n_3 b, \ m_3 = m_1 - m_2 q_2, n_3 = n_1 - n_2 q_2 \in \mathbb{Z}$$

In general, we obtain $r_k = m_{k+1} a + n_{k+1} b = ma + nb$ where $m = m_{k+1}$, $n = n_{k+1} \in \mathbb{Z}$, hence $d = ma + nb$. $\qquad\square$

Example 3.12 Calculate the greatest common divisor d of following pairs of numbers x, y. Also, find numbers u, v (Bézout coefficients) such that $ux + vy = d$.

1. $d_1 = (141, 96)$. Applying the Euclidean algorithm:

$$141 = 96(1) + 45$$
$$96 = 45(2) + 6$$
$$45 = 6(7) + 3$$
$$6 = 3(2) + 0$$

Hence $d_1 = (141, 96) = 3$. Indeed, $191 = 3 \cdot 47$ and $96 = 3 \cdot 32$. On the other hand:

$$3 = 45 - 7(6)$$
$$= 45 - 7[96 - 45(2)]$$
$$= 15(45) - 7(96)$$
$$= 15[141 - 96(1)] - 7(96)$$
$$= 141(15) + 96(-22)$$

So that the Bézout coefficients are $u = 15, v = -22$ and $141(15) + 96(-22) = 2115 - 2112 = 3$.

2. $d_2 = (-2805, 119) = 17$. Indeed, applying the Euclidean algorithm:

$$-2805 = 119(-24) + 51$$
$$119 = 51(2) + 17$$
$$51 = 17(3) + 0$$

In addition, it is not hard to obtain $17 = 119(-47) - 2805(-2)$.

3. $d_3 = (726, 275) = 11$ and $11 = 726(11) + 275(-29)$. Arithmetic operations are left to the reader as an exercise.

Proposition 3.3 *If $a \mid c$, $b \mid c$ and $(a, b) = d$ then $ab \mid cd$.*

Proof By hypothesis, there are integers $m, n \in \mathbb{Z}$ such that $c = am$ and $c = bn$. By the Bezout's identity $d = pa + qb$ where $p, q \in \mathbb{Z}$, then

$$cd = c(pa + qb) = cpa + cqb = bnpa + amqb = ab(np + mq), \ np + mq \in \mathbb{Z}$$

Hence $ab \mid cd$. □

Definition 3.18 Let $a, b \in \mathbb{Z}$ no both zero. It is said that a and b are **relatively prime** (mutually prime or coprime) if the only positive integer that evenly divides both of them is 1, i.e., $(a, b) = 1$.

Theorem 3.8 *Let $a, b \in \mathbb{Z}$.*

1. *If $1 = (a, b)$ and $a \mid bc$ then $a \mid c$.*
2. *If $1 = (a, b)$, then $1 = (a, b^n) \ \forall n \in \mathbb{N}$*

Proof 1. Since $(a, b) = 1$ then there exist $m, n \in \mathbb{Z}$ such that $am + bn = 1$, so that $c = a(cm) + bc(n)$. On the other hand, since $a \mid a$ and $a \mid bc$ then $a \mid (a(cm) + bc(n))$. Therefore $a \mid c$.

2. Let A be the set of positive integers that satisfy the statement to prove. If $1 = (a, b)$, then $1 = (a, b^1)$ so that $1 \in A$. Suppose that $n = k \in A$, i.e., if $1 = (a, b)$, then $1 = (a, b^k)$, hence there exist $m, n \in \mathbb{Z}$ such that $1 = am + b^k n$. On the other hand, since $(a, b) = 1$ then there exist $x, y \in \mathbb{Z}$ such that $1 = ax + by$, so that

$$1 = 1 \cdot 1 = (ax + by)(am + b^k n) = a(axm + bym + xb^k n) + b^{k+1}(ny)$$

Therefore $1 = (a, b^{k+1})$, i.e., $k + 1 \in A$ and $A = \mathbb{N}$. Based in the mathematical induction principle, if $1 = (a, b)$, then $1 = (a, b^n) \ \forall n \in \mathbb{N}$. □

Definition 3.19 It is said that $p \in \mathbb{Z}$ is a **prime number**, if $p > 1$ and the only positive divisors of p are 1 and itself.

Theorem 3.9 *Let $a, b \in \mathbb{Z}$ and let p a prime number. Then*

1. *$p \mid a$ or $1 = (a, p)$.*
2. ***Euclid's lemma.*** *If $p \mid ab$ then $p \mid a$ or $p \mid b$.*

Proof 1. Let $d = (a, p)$ then $d \mid a$ and $d \mid p$. Since p is prime then $d = p$ or $d = 1$.
If $d = p$ then $p \mid a$. On the other hand if $d = 1$ it follows that $1 \mid a$ and $1 \mid p$
then $1 = (a, p)$.
2. Suppose $p \nmid a$ then $1 = (p, a)$. By hypotheses $p \mid ab$ then by **Theorem** 3.8 $p \mid b$.
In the same manner, it is proved that $p \mid a$. $\square$

Lemma 3.6 *If $a \in \mathbb{Z}$ and $a > 1$, then the smallest integer greater than 1 and that divides a, is a prime number.*

Proof Let $B = \{m \in \mathbb{N} \mid m > 1$ and $m \mid a\}$. Since $a > 1$ and $a \mid a$ then $a \in B$, so that $B \neq \emptyset$. By the well-ordering principle, there exists $p \in B$ such that $p \leq m \, \forall \, m \in B$. If p is not prime then there exists $q \in \mathbb{Z}$ with $1 < q < p$ such that $q \mid p$. Since $p \mid a$ then $q \mid a$ so that $q \in B$ which is a contradiction. Therefore p is prime. $\square$

Theorem 3.10 (Fundamental theorem of arithmetic) *If $a \in \mathbb{Z}$ and $a > 1$, then a is prime or there exist $p_1, p_2, \ldots, p_k$ prime numbers such that:*

$$a = p_1 \cdot p_2 \cdot \ldots \cdot p_k$$

Furthermore, if $q_1, q_2, \ldots, q_m$ are prime numbers such that

$$a = q_1 \cdot q_2 \cdot \ldots \cdot q_m$$

then $m = k$ and $q_i = p_j$ for some $i, j = 1, 2, \ldots, k$.

Proof The proof is left to the reader as an exercise. $\square$

Example 3.13 Let $a = 2241756$, $b = 8566074$. The unique factorization of both numbers is:

$$2241756 = 2^2 \cdot 3^4 \cdot 11 \cdot 17 \cdot 37$$
$$8566074 = 2 \cdot 3^4 \cdot 11^2 \cdot 19 \cdot 23$$

In addition, $(a, b) = 2 \cdot 3^4 \cdot 11$.

3.4.4 Least Common Multiple

All the integers numbers $a_1, a_2, \ldots, a_n$ different from zero have a common multiple m if $a_i \mid b \, \forall 1 \leq i \leq n$. A common multiple of $a_1, a_2, \ldots, a_n$ is

$$m = a_1 \cdot a_2 \cdots a_n = \prod_{i=1}^{n} a_i$$

since

$$m = a_1(a_2 \cdots a_n)$$
$$m = a_2(a_1 \cdots a_n)$$
$$\vdots$$
$$m = a_n(a_1 \cdot a_2 \cdots a_{n-1})$$

It should be mentioned that there may be several common multiples, but definitely, there is one minimum multiple.

Definition 3.20 Let $A = \{a_1, a_2, a_3, \ldots, a_n\}$ a set of integers. The **least common multiple** (ℓmc) is the smallest positive integer that is divisible by all the numbers $a_i \in A$, $1 \le i \le n$. This number is denoted by

$$\ell \, mc = [a_1, a_2, a_3, \ldots, a_n]$$

Theorem 3.11 *If b is any common multiple of* $a_1, a_2, a_3, \ldots, a_n$, *then* $\ell \, mc \mid b$.

Proof Since b is any multiple of $a_1, a_2, a_3, \ldots, a_n$ then $a_i \mid b \, \forall 1 \le i \le n$. Let $h = [a_1, a_2, a_3, \ldots, a_n]$. Applying the division algorithm to h and b: $b = h \cdot q + r$, $0 \le r < |h|$. Let suppose that $r \ne 0$ then $r > 0$ or $r < 0$. On the other hand, we know that $a_i \mid b \, \forall 1 \le i \le n$ and $a_i \mid h$ then $a_i \mid r$, where $r = b + h(-q)$, so that r is a multiple of $a_i < n$. This is a contradiction to the division algorithm and the fact that h is the least common multiple. Therefore $r = 0$, $b = h \cdot q$ and $h \mid b$. □

Theorem 3.12 *Let* $m > 0$, *then*

1. $[ma, mb] = m[a, b]$
2. $[a, b](a, b) = |a \cdot b|$

Proof 1. Let $h = [ma, mb]$ then $ma \mid h$ and $mb \mid h$ so that there are integers q_1, q_2 such that $h = (ma)q_1 = m(aq_1) = a(mq_1)$ and $h = (mb)q_2 = m(bq_2) = b(mq_2)$. It is clear that $m \mid h$, $a \mid h$ and $b \mid h$ then $h = m \cdot h_1$. Let $h_2 = [a, b]$ then $a \mid h_2$ and $b \mid h_2$ so that $ma \mid mh_2$ and $mb \mid mh_2$ where mh_2 is another common multiple of ma and mb. This implies that $h \mid mh_2$ and $mh \mid mh_2$ so that $h_1 \mid h_2$. On the other hand $ma \mid mh_1$ and $mb \mid mh_1$ then $a \mid h_1$ and $b \mid h_2$, hence $h_2 \mid h_1$. Therefore $h_1 = h_2$ and $[ma, mb] = m[a, b]$.
2. The proof of this item is left to reader as an exercise. □

Example 3.14 According to a previous example, $(141, 96) = 3$. From the previous theorem, the least common multiple of these numbers is:

$$[141, 96] = \frac{|141 \cdot 96|}{(141, 96)} = \frac{2^5 \cdot 3^2 \cdot 47}{3} = 4512$$

3.5 Polynomials Rings

Definition 3.21 The **polynomial ring**, $\mathbb{F}[x]$ in x over a field $\mathbb{F}$ is defined as the set of **polynomials** $f(x)$, i.e.,

$$\mathbb{F}[x] = \{f(x) = a_0 x_0 + a_1 x_1 + \cdots + a_n x_n \mid n \in \mathbb{Z}, n \geq 0, a_i \in \mathbb{F}\}$$

where $a_0, a_1, \ldots, a_n$ are the **coefficients** of $p(x)$.

Definition 3.22 Let $f(x) = a_0 + a_1 x + \cdots + a_n x^n \in \mathbb{F}[x]$, $a_n \neq 0$. The **degree** of a polynomial $f(x)$, written d$^\circ f(x)$ is the largest n such that the coefficient of x^n is not zero. In this case the coefficient a_n is called the **leading coefficient**. If $a_n = 1$, $f(x)$ is called monic.

Example 3.15 Consider $\mathbb{F} = \mathbb{R}$, let $f(x) = x^6 - 7x^4 + 8x^2 + 2x - 7$, $g(x) = 2x^3 + 5$ then $f(x), g(x) \in \mathbb{R}[x]$ with d$^\circ f(x) = 6$ and d$^\circ g(x) = 3$.

Definition 3.23 Let $f(x), g(x) \in \mathbb{F}[x]$, where $f(x) = a_0 + a_1 x + \cdots + a_n x^n$ and $g(x) = b_0 + b_1 x + \cdots + b_n x^n$. It is said that $f(x)$ equals $g(x)$ if $a_i = b_i$ $\forall i = 0, 1, 2, \ldots, n$.

Definition 3.24 Let $f(x), g(x) \in \mathbb{F}[x]$, where

$$f(x) = a_0 + a_1 x + \cdots + a_n x^n$$

and

$$g(x) = b_0 + b_1 x + \cdots + b_m x^m$$

- The **sum** of polynomials $f(x)$ and $g(x)$ is the polynomial given by:

$$f(x) + g(x) = (a_0 + b_0) + (a_1 + b_1) x + \cdots + (a_s + b_s) x^s$$

where $s = \max\{n, m\}$.
- The **product** of polynomials $f(x)$ and $g(x)$ is the polynomial given by:

$$f(x) \cdot g(x) = c_0 + c_1 x + \cdots + c_l x^l$$

where $l = m + n$, $c_k = \displaystyle\sum_{i+j=k} a_i b_j = a_0 b_k + a_1 b_{k-1} + \cdots + a_k b_0$, $k = 0, 1, \ldots,$ $n + m$, $i = 0, 1, \ldots, n$ and $j = 0, 1, \ldots, m$.

Exercise 3.4 $\mathbb{F}[x]$ is a commutative ring with identity. The proof is left to the reader as an exercise.

Example 3.16 Let $p(x) = ax^2 + bx + c$. Calculate the values of a, b, c so that the next equality holds:

$$x^3 - 17x^2 + 54x - 8 = (x - 4)\, p(x)$$

According to the product of polynomials defined above, we have that

$$(x - 4)\, p(x) = ax^3 + (b - 4a)\, x^2 + (c - 4b)\, x - 4c$$

then, by **Definition** 3.23:

$$a = 1$$
$$b - 4a = -17$$
$$c - 4b = 54$$
$$-4c = -8$$

Finally, we find $a = 1, b = 13, c = 2$ and $p(x) = x^2 - 13x + 2$.

After performing operations between polynomials, it is common to ask what happens to the degree of the resulting polynomial. The following proposition establishes what is the degree of this polynomial.

Proposition 3.4 *Let $f(x), g(x) \in \mathbb{F}[x]$ with $f(x) \neq 0$ and $g(x) \neq 0$, then:*

1. $d° (f(x) + g(x)) \leq \max \{d° f(x), d° g(x)\}$
2. $d° (f(x) \cdot g(x)) = d° f(x) + d° g(x)$

Proof Let $f(x) = a_0 + a_1 x + \cdots + a_n x^n$ and $g(x) = b_0 + b_1 x + \cdots + b_m x^m$. Since $f(x) \neq 0$ and $g(x) \neq 0$, then $a_n \neq 0$ and $b_m \neq 0$, so that $d° f(x) = n$ and $d° g(x) = m$.

1. Consider that $m > n$ or $m < n$. Suppose that $f(x) + g(x) \neq 0$ then

$$f(x) + g(x) = (a_0 + b_0) + (a_1 + b_1) x + \cdots + (a_s + b_s) x^s$$

where $s = \max \{n, m\}$, hence

$$d° (f(x) + g(x)) = s = \max \{d° f(x), d° g(x)\}$$

On the other hand, if $m = n$ then

$$f(x) + g(x) = (a_0 + b_0) + (a_1 + b_1) x + \cdots + (a_s + b_s) x^s$$

where $a_j + b_j \neq 0$ for some $j = 0, 1, \ldots, s$. Let $k \in \{0, 1, \ldots, s\}$ the max such that $a_k + b_k \neq 0$ then

$$d° (f(x) + g(x)) = k \leq s = \max \{d° f(x), d° g(x)\}$$

2. By definition, $f(x) \cdot g(x) = c_0 + c_1 x + \cdots + c_l x^l$, where $c_k = \displaystyle\sum_{i+j=k} a_i b_j$ and $l = m + n$. It is clear that $c_l \neq 0$ so that $f(x) \cdot g(x) \neq 0$, then $\mathrm{d}^\circ (f(x) \cdot g(x)) = l = n + m = \mathrm{d}^\circ f(x) + \mathrm{d}^\circ g(x)$. $\qquad\square$

Exercise 3.5 Let $f(x), g(x) \in \mathbb{F}[x]$ with $f(x) \neq 0$ and $g(x) \neq 0$, $c \in \mathbb{F}, c \neq 0$. Prove that:

1. $\mathrm{d}^\circ f(x) \leq \mathrm{d}^\circ (f(x) \cdot g(x))$.
2. $\mathrm{d}^\circ cf(x) = \mathrm{d}^\circ f(x)$.
3. $\mathrm{d}^\circ (f(x) + c) = \mathrm{d}^\circ f(x)$.

Taking as a starting point the division algorithm for integers, the following theorem states that given $f(x), g(x) \in \mathbb{F}[x], g(x) \neq 0$, there are only $q(x), r(x) \in \mathbb{F}[x]$ such that $f(x) = q(x)g(x) + r(x)$ where $\mathrm{d}^\circ r(x) < \mathrm{d}^\circ g(x)$ or $r(x) = 0$.

Theorem 3.13 (Division Algorithm for Polynomials) *If $f(x), g(x) \in \mathbb{F}[x], g(x) \neq 0$, there are only $q(x), r(x) \in \mathbb{F}[x]$ such that*

$$f(x) = g(x)q(x) + r(x)$$

where $r(x) = 0$ or $\mathrm{d}^\circ r(x) < \mathrm{d}^\circ g(x)$. The polynomials $q(x)$ and $r(x)$ are called quotient and residue respectively.

Example 3.17 If the polynomial $x^5 - 5x^4 + 10x^3 - 10x^2 + 5x + 9$ is divided by $x - 1$, to find the quotient $q(x)$ and residue $r(x)$, we consider the following arrangement:

$$
\begin{array}{r}
x^4 - 4x^3 + 6x^2 - 4x + 1 \\
\hline
x - 1)\, \overline{\ x^5 - 5x^4 + 10x^3 - 10x^2 + 5x + 9} \\
-x^5 + x^4 \\
\hline
-4x^4 + 10x^3 \\
4x^4 - 4x^3 \\
\hline
6x^3 - 10x^2 \\
-6x^3 + 6x^2 \\
\hline
-4x^2 + 5x \\
4x^2 - 4x \\
\hline
x + 9 \\
-x + 1 \\
\hline
10
\end{array}
$$

Therefore $q(x) = x^4 - 4x^3 + 6x^2 - 4x + 1$ and $r(x) = 10$. According to the division algorithm for polynomials:

$$x^5 - 5x^4 + 10x^3 - 10x^2 + 5x + 9 = (x - 1)\left(x^4 - 4x^3 + 6x^2 - 4x + 1\right) + 10$$

Similar to the division algorithm for polynomials, the greatest common divisor for polynomials can be defined and the Euclidean algorithm can be used.

Example 3.18 Let $f(x) = 16x^3 + 36x^2 - 12x - 18$ and $g(x) = 8x^2 - 2x - 3$. The greatest common divisor of $f(x)$ and $g(x)$ is obtained in the same manner as for integer numbers:

$$16x^3 + 36x^2 - 12x - 18 = (8x^2 - 2x - 3)(2x + 5) + 4x - 3$$
$$8x^2 - 2x - 3 = (4x - 3)(2x + 1) + 0$$

hence $\gcd(f(x), g(x)) = 4x - 3$, where:

$$
\begin{array}{r}
2x + 5 \\
\hline
8x^2 - 2x - 3 \overline{)\; 16x^3 + 36x^2 - 12x - 18} \\
-16x^3 + 4x^2 + 6x \\
\hline
40x^2 - 6x - 18 \\
-40x^2 + 10x + 15 \\
\hline
4x - 3
\end{array}
\qquad
\begin{array}{r}
2x + 1 \\
\hline
4x - 3 \overline{)\; 8x^2 - 2x - 3} \\
-8x^2 + 6x \\
\hline
4x - 3 \\
-4x + 3 \\
\hline
0
\end{array}
$$

On the other hand, it is left as an exercise to the reader, verify that:

$$\gcd\left(2x^3 - 11x^2 + 10x + 8,\, 2x^3 + x^2 - 8x - 4\right) = 2x^2 - 3x - 2.$$

References

1. Birkhoff, G., Mac Lane, S.: A Survey of Modern Algebra. Universities Press (1965)
2. Dummit, D.S., Foote, R.M.: Abstract Algebra, vol. 3. Hoboken, Wiley (2004)
3. Fraleigh, J.B.: A First Course in Abstract Algebra. Addison-Wesley (2003)
4. Lang, S.: Algebra. Springer (2002)
5. Saracino, D.: Abstract Algebra: A First Course. Waveland Press (2008)
6. Herstein, I.N.: Abstract Algebra. Prentice Hall (1996)
7. Hungerford, T.W.: Abstract Algebra: An Introduction. Cengage Learning (2012)
8. Jacobson, N.: Lectures in Abstract Algebra, vol. 1. Van Nostrand, Basic concepts (1951)
9. Kunz, E.: Introduction to Commutative Algebra and Algebraic Geometry. Birkhuser (1985)
10. Samuel, P., Serge, B.: Mthodes d'algbre abstraite en gomtrie algbrique. Springer (1967)
11. Herstein, I.N.: Topics in Algebra. Wiley (2006)
12. Bourbaki, N.: Commutative Algebra, vol. 8. Hermann, Paris (1972)
13. Zariski, O., Samuel, P.: Commutative Algebra, vol. 1. Springer Science & Business Media (1958)
14. Zariski, O., Samuel, P.: Commutative Algebra, vol. 2. Springer Science & Business Media (1960)

Chapter 4
Matrices and Linear Equations Systems

Abstract In this chapter basic concepts about matrices and linear algebra is introduced, in order to present the information in a easy to understand manner, the first subject is algebraic operations with real numbers, then linear operation with real number vectors, followed by the definition of a matrix and its basic operations, next the change of basis for a matrix is given, finally Gauss Jordan method is presented.

4.1 Properties of Algebraic Operations with Real Numbers

The following list describes the **properties of the addition** of real numbers.

1. **Associative property**. For every real numbers α, β, γ, $(\alpha + \beta) + \gamma = \alpha + (\beta + \gamma)$.
2. **Commutative property**. For every real numbers α, β, $\alpha + \beta = \beta + \alpha$.
3. **Existence of an additive identity element**. There exist a real number γ such that for every $\alpha \in \mathbb{R}$, $\alpha + \gamma = \alpha$. This number is unique and it is called the *zero number* (0).
4. **Existence of an additive inverse**. For every real number α there exists $\beta \in \mathbb{R}$ such that $\alpha + \beta = 0$. It is easy to show that for every number $\alpha \in \mathbb{R}$, the number β with the above property is unique. This number is called *additive inverse* of α and $\beta = -\alpha$.

For example, if $a + c = b + c$ then $a = b$. To demonstrate this assertion, consider the above properties. According to additive identity property 3, $a = a + 0$. On the other hand, $a + 0 = a + (c + (-c))$ considering the existence of additive inverse property 4. Continuing this argumentation we have the following:

$$a \underset{3}{=} a + 0 \underset{4}{=} a + (c + (-c)) \underset{1}{=} (a + c) + (-c) \underset{\text{By hypothesis}}{=} (b + c) + (-c)$$

$$(b + c) + (-c) \underset{1}{=} b + (c + (-c)) \underset{4}{=} b + 0 \underset{3}{=} b$$

© Springer Nature Switzerland AG 2019

R. Martínez-Guerra et al., *Algebraic and Differential Methods for Nonlinear Control Theory*, Mathematical and Analytical Techniques with Applications to Engineering, https://doi.org/10.1007/978-3-030-12025-2_4

The **properties of the multiplication** of real numbers are listed below.

5. **Associative property**. For every real numbers α, β, γ, $(\alpha\beta)\,\gamma = \alpha\,(\beta\gamma)$.
6. **Commutative property**. For every real numbers α, β, $\alpha\beta = \beta\alpha$.
7. **Existence of a multiplicative identity element**. There exist a real number δ such that for every $\alpha \in \mathbb{R}$, $\alpha\delta = \alpha$. This number is unique and it is called the *one number* (1).
8. **Existence of a multiplicative inverse**. For every real number $\alpha \neq 0$ there exists $\beta \in \mathbb{R}$ such that $\alpha\beta = 1$. It is easy to show that for every number $\alpha \in \mathbb{R}$, the number β with the above property is unique. This number is called *multiplicative inverse* of α and $\beta = \alpha^{-1}$.

Finally the distributive property of multiplication respect to addition that describes the multiplication of a sum by a factor, where each summand is multiplied by this factor and the resulting products are added:

9. **Distributive property**. For every real numbers α, β, γ, $\alpha\,(\beta + \gamma) = \alpha\beta + \alpha\gamma$ (left) and $(\beta + \gamma)\,\alpha = \beta\alpha + \gamma\alpha$ (right).

In addition, $1 \neq 0$. According to these properties, the set of real numbers $\mathbb{R}$ is called a field. A field $\mathbb{F}$ is a set together with two operations, usually called addition and multiplication, and denoted by $+$ and $\cdot$, respectively, such that the above properties (axioms) hold. The rational numbers $\mathbb{Q}$ and complex numbers $\mathbb{C}$ are important examples of fields.

4.2   The Set $\mathbb{R}^n$ and Linear Operations

A **list** of **length** n is an ordered collection of n objects. The of n objects can be numbers or abstract entities, for example:

$$(x_1, \ldots, x_n)$$

In particular, when $n = 2$, the list of length 2 is called an ordered pair or if $n = 3$, the list is called ordered triple [1–3].

Definition 4.1 A list of n real numbers is called **n-tuple** of real numbers. If x is a n-tuple of real numbers then its entries are $x_1, x_2, \ldots, x_n$. This n-tuple can be write as follows:

$$x = \left[x_j\right]_{j=1}^n$$

If $j \in \{1, \ldots, n\}$ then x_j is the j-th component of x.

Considering the above definition, we can establish the condition for the tuples equality. Let x a n-tuple and y a m-tuple, then x and y are **equal** if and only if $m = n$ and $x_j = y_j \; \forall \, j \in \{1, \ldots, n\}$.

Definition 4.2 The set of every n-tuples of real numbers is called the set $\mathbb{R}^n$.

4.2.1 Linear Operations in $\mathbb{R}^n$

The **addition** of n-tuples and **multiplication** of a number by a n-tuples are defined by components, that is, element by element. Let $x, y \in \mathbb{R}^n$, and $\alpha \in \mathbb{R}$. The addition of 2 n-tuples is given by

$$x + y = \left[x_j\right]_{j=1}^n + \left[y_j\right]_{j=1}^n = \left[x_j + y_j\right]_{j=1}^n$$

where $x + y$ is an element of $\mathbb{R}^n$ whose j-th component is $(x + y)_j = x_j + y_j$ for $j \in \{1, \ldots, n\}$. On the other hand, the multiplication of a n-tuple by one number is defined as follows:

$$\alpha x = \alpha \left[x_j\right]_{j=1}^n = \left[\alpha x_j\right]_{j=1}^n$$

where αx is an element of $\mathbb{R}^n$ whose j-th component is $(\alpha x)_j = \alpha x_j$ for $j \in \{1, \ldots, n\}$.

Example 4.1 Consider the case $n = 4$. Let $x, y \in \mathbb{R}^4$ given by

$$x = \begin{bmatrix} \pi \\ \sqrt[5]{2} \\ e \\ 2 \end{bmatrix}, \quad y = \begin{bmatrix} \sqrt{\pi} \\ 8 \\ e^5 \\ 7\pi \end{bmatrix}$$

and $\alpha = 2$, then

$$x + y = \begin{bmatrix} \pi \\ \sqrt[5]{2} \\ e \\ 2 \end{bmatrix} + \begin{bmatrix} \sqrt{\pi} \\ 8 \\ e^5 \\ 7\pi \end{bmatrix} = \begin{bmatrix} \pi + \sqrt{\pi} \\ 8 + \sqrt[5]{2} \\ e + e^5 \\ 2 + 7\pi \end{bmatrix}, \quad \alpha y = 2 \begin{bmatrix} \sqrt{\pi} \\ 8 \\ e^5 \\ 7\pi \end{bmatrix} = \begin{bmatrix} 2\sqrt{\pi} \\ 16 \\ 2e^5 \\ 14\pi \end{bmatrix}$$

Remark 4.1 The n-tuples of real numbers that satisfy the above operations, are called **vectors** and the numbers that are multiply by these vectors usually are called **scalars**.

The properties of real numbers as associativity and commutativity of addition, they are also satisfied by the tuples. Similarly, a zero tuple and an opposite tuple (seen as the additive inverse) are defined. Then, for the addition in $\mathbb{R}^n$:

1. **Associative property of $\mathbb{R}^n$** For every $x, y, z \in \mathbb{R}^n$ $(x + y) + z = x + (y + z)$.

2. **Commutative property of $\mathbb{R}^n$** For every $x, y \in \mathbb{R}^n$ $x + y = y + x$.

It is easy to check the properties 1 and 2. For the first one,

$$(x + y) + z \quad \underbrace{=}_{\text{By notation}} \quad \left([x_j]_{j=1}^n + [y_j]_{j=1}^n \right) + [z_j]_{j=1}^n$$

$$\underbrace{=}_{\text{Definition of sum in } \mathbb{R}^n} \quad [x_j + y_j]_{j=1}^n + [z_j]_{j=1}^n$$

$$\underbrace{=}_{\text{Definition of sum in } \mathbb{R}^n} \quad [(x_j + y_j) + z_j]_{j=1}^n$$

$$\underbrace{=}_{\text{Associative property 1 for } \mathbb{R}} \quad [x_j + (y_j + z_j)]_{j=1}^n$$

$$\underbrace{=}_{\text{Definition of sum in } \mathbb{R}^n} \quad [x_j]_{j=1}^n + [y_j + z_j]_{j=1}^n$$

$$\underbrace{=}_{\text{Definition of sum in } \mathbb{R}^n} \quad [x_j]_{j=1}^n + \left([y_j]_{j=1}^n + [z_j]_{j=1}^n \right)$$

$$\underbrace{=}_{\text{By notation}} \quad x + (y + z)$$

On the other hand, to verify the property 2, first one we check that the tuples $x + y$ and $y + x$ have the same length. Indeed, for the definition of addition in $\mathbb{R}^n$ we conclude that both tuples have length n. Let an index $k \in \{1, \ldots, n\}$ and show that the $k - $ th component of $x + y$ is equal to the $k - $ th component of $y + x$:

$$(x + y)_k \quad \underbrace{=}_{\text{Definition of sum in } \mathbb{R}^n} \quad x_k + y_k$$

$$\underbrace{=}_{\text{Commutative property 2 for } \mathbb{R}} \quad y_k + x_k$$

$$\underbrace{=}_{\text{Definition of sum in } \mathbb{R}^n} \quad (y + x)_k$$

We have shown that the corresponding components of the tuples $x + y$ and $y + x$ are equal.

The n-tuple $[0]_{k=1}^n$, denoted by $\mathbf{0}_n$ is called the **nulln-tuple**. This tuple has length n and every component of this tuple is zero, that is, $\mathbf{0}_n \in \mathbb{R}^n \ \forall k \in \{1, \ldots, n\}$. Equivalently $(\mathbf{0}_n)_k = 0$. This tuple is the neutral element (identity element) of the sum in $\mathbb{R}^n$, that is, $x + \mathbf{0}_n = \mathbf{0}_n + x = x \ \forall x \in \mathbb{R}^n$. To verify this assertion, it is sufficient to verify the first (or the second) equality because the sum is commutative in $\mathbb{R}^n$.

Let $x = [x_k]_{k=1}^n$ then by notation and definition of the null n-tuple, we have: $x + \mathbf{0}_n = [x_k]_{k=1}^n + [0]_{k=1}^n$. Now, according to definition of sum in $\mathbb{R}^n$, $[x_k]_{k=1}^n + [0]_{k=1}^n = [x_k + 0]_{k=1}^n$ and finally by the additive identity element 3 and definition of x, $[x_k + 0]_{k=1}^n = [x_k]_{k=1}^n = x$.

On the other hand, the element in $\mathbb{R}^n$ there exist an element $-x \in \mathbb{R}^n$ such that $x + (-x) = \mathbf{0}_n$. This element is unique and is called the opposite tuple or **additive inverse** in $\mathbb{R}^n$ of x. It is easy to check that $x + (-x) = (-x) + x = \mathbf{0}_n$ for every $x \in \mathbb{R}^n$. Finally, it is left as an exercise to demonstrate the following properties:

1. **Distributive property of multiplication by scalars in $\mathbb{R}^n$ respect to addition of scalars**. For every $\alpha, \beta \in \mathbb{R}$ and $x \in \mathbb{R}^n$, $(\alpha + \beta) x = \alpha x + \beta x$.

2. **Distributive property of multiplication by scalars in $\mathbb{R}^n$ respect to addition of vectors**. For every $\alpha \in \mathbb{R}$ and $x, y \in \mathbb{R}^n$, $\alpha(x + y) = \alpha x + \alpha y$.
3. For every $x \in \mathbb{R}^n$, $1 \in \mathbb{R}$, $1x = x$.
4. For every $\alpha, \beta \in \mathbb{R}$ and $x \in \mathbb{R}^n$, $\alpha(\beta x) = (\alpha\beta)x$.

4.3 Background of Matrix Operations

Letx $f : \mathbb{R}^4 \to \mathbb{R}^2$ defined by

$$f : \begin{bmatrix} x_1 \\ x_2 \\ x_3 \\ x_4 \end{bmatrix} \longmapsto \begin{bmatrix} y_1 \\ y_2 \end{bmatrix}$$

where $y_1 = 3x_2 + 5x_3$, $y_2 = 3x_1 - 4x_2 + 5x_3$. Later, it will define what is a matrix and a linear transformation and the relationship between them. For now, we will say that a matrix is used to describe a linear transformation. According to function f [4, 5].

Definition 4.3 Let $\mathbb{F}$ a field. Let $m, n \in \mathbb{N}$. A **matrix** $m \times n$ with entries in $\mathbb{F}$ is a rectangular array with m rows and n columns with elements of $\mathbb{F}$:

$$A = \begin{pmatrix} a_{11} & a_{12} & \cdots & a_{1n} \\ a_{21} & a_{22} & \cdots & a_{2n} \\ \vdots & \vdots & \ddots & \vdots \\ a_{m1} & a_{m2} & \cdots & a_{mn} \end{pmatrix}$$

More strictly, an $m \times n$ matrix over the field $\mathbb{F}$ is a function A from the set of pairs of integers (i, j), $1 \le i \le m$, $1 \le j \le n$, into the field $\mathbb{F}$. The entries of the matrix A are the scalars $a_{i,j}$. Therefore, a matrix can be represented by $A = (a_{ij})_{i,j}$.

Note that in later chapters the entries of matrices can be different than elements of fields, so they can be elements of rings or even other matrices, which are more general sets and with specific features.

Definition 4.4 Let $A = (a_{ij})_{i,j}$, $B = (b_{ij})_{i,j} \in \mathscr{M}_{m \times n}(\mathbb{F})$. Let $c \in \mathbb{F}$. The **sum (or addition) of matrices** is given by [1, 2, 6, 7]:

$$A + B = (a_{ij} + b_{ij})_{i,j}$$

The **multiplication of one matrix by a scalar** c is given by:

$$c \cdot A = (c \cdot a_{ij})_{i,j}$$

Definition 4.5 Let $A = \left(a_{ij}\right)_{i,j} \in \mathcal{M}_{m \times n}(\mathbb{F})$, $B = \left(b_{jk}\right)_{j,k} \in \mathcal{M}_{n \times p}(\mathbb{F})$. The **product** AB is the matrix $C = (c_{ik})_{i,k} \in \mathcal{M}_{m \times p}(\mathbb{F})$ such that [7–9]:

$$c_{ik} = \sum_{j=1}^{n} a_{ij} b_{jk}$$

Proposition 4.1 *Let* $A = \left(a_{ij}\right)_{i,j} \in \mathcal{M}_{m \times n}(\mathbb{F})$, $B = \left(b_{jk}\right)_{j,k} \in \mathcal{M}_{n \times p}(\mathbb{F})$, $C = (c_{kl})_{k,l} \in \mathcal{M}_{p \times q}(\mathbb{F})$. *Then:*

1. $A(BC) = (AB)C$
2. $A(B + C) = AB + AC$
3. $(A + B)C = AC + BC$
4. $k(AB) = (kA)B$

Proof 1. Let $A = \left[a_{ij}\right] \in \mathcal{M}_{m \times n}$, $B = \left[b_{jk}\right] \in \mathcal{M}_{n \times p}$ and $C = [c_{kl}] \in \mathcal{M}_{p \times q}$, defining $\left[\alpha_{jl}\right] = BC = \sum_{k=1}^{p} b_{jk} c_{kl}$ and $[\beta_{ik}] = AB = \sum_{j=1}^{n} a_{ij} b_{jk}$ then:

$$\gamma_{il} = A\,(BC)$$
$$= A\alpha_{jl}$$
$$= A\left(\sum_{k=1}^{p} b_{jk} c_{kl}\right)$$
$$= \sum_{j=1}^{n} a_{ij} \left(\sum_{k=1}^{p} b_{jk} c_{kl}\right)$$
$$= \sum_{j=1}^{n} \sum_{k=1}^{p} a_{ij} \left(b_{jk} c_{kl}\right)$$

Also

$$\delta_{il} = (AB)\,C$$
$$= \beta_{ik} C$$
$$= \left(\sum_{j=1}^{n} a_{ij} b_{jk}\right) C$$
$$= \sum_{k=1}^{p} \left(\sum_{j=1}^{n} a_{ij} b_{jk}\right) c_{kl}$$
$$= \sum_{k=1}^{p} \sum_{j=1}^{n} \left(a_{ij} b_{jk}\right) c_{kl}$$

Since $\sum_{j=1}^{n} \sum_{k=1}^{p} a_{ij} \left(b_{jk} c_{kl} \right) = \sum_{k=1}^{p} \sum_{j=1}^{n} \left(a_{ij} b_{jk} \right) c_{kl}$ it is concluded that $A(BC) = (AB)C$.

2. Let $A = \left[a_{ij} \right] \in \mathcal{M}_{m \times n}$, $B = \left[b_{jk} \right] \in \mathcal{M}_{n \times p}$ and $C = \left[c_{jk} \right] \in \mathcal{M}_{n \times p}$, by defining $D = \left[d_{jk} \right] = B + C = b_{jk} + c_{jk}$, it is easy to see that the matrix $A(B + C)$ equals

$$AD = \sum_{j=1}^{n} a_{ij} d_{jk}$$

$$= \sum_{j=1}^{n} a_{ij} \left(b_{jk} + c_{jk} \right)$$

And $AB + AC$ is

$$AB + AC = \sum_{j=1}^{n} a_{ij} b_{jk} + \sum_{j=1}^{n} a_{ij} c_{jk}$$

$$= \sum_{j=1}^{n} a_{ij} \left(b_{jk} + c_{jk} \right)$$

Thus $AD = AB + AC$ showing that $A(B + C) = AB + BC$ is true.

3. Having $A = \left[a_{ij} \right] \in \mathcal{M}_{m \times n}$, $B = \left[b_{ij} \right] \in \mathcal{M}_{m \times n}$ and $C = \left[c_{jk} \right] \in \mathcal{M}_{n \times p}$, and making $D = \left[d_{ij} \right] = A + B = a_{ij} + b_{ij}$ causes that:

$$(A + B)C = DC$$

$$= \sum_{j=1}^{n} d_{ij} c_{jk}$$

$$= \sum_{j=1}^{n} \left(a_{ij} + b_{ij} \right) c_{jk}$$

$$\sum_{j=1}^{n} \left(a_{ij} c_{jk} + b_{ij} c_{jk} \right)$$

Also $AC + BC$ cause:

$$AC + BC = \sum_{j=1}^{n} a_{ij} c_{jk} + \sum_{j=1}^{n} b_{ij} c_{jk}$$

$$= \sum_{j=1}^{n} \left(a_{ij} c_{jk} + b_{ij} c_{jk} \right)$$

Hence $(A + B)C = AC + BC$

4. With $A = \left[a_{ij}\right] \in \mathscr{M}_{m \times n}$, $B = \left[b_{jl}\right] \in \mathscr{M}_{n \times p}$ and $k \in \mathbb{F}$ it can be seen that:

$$k\left(AB\right) = k\left(\sum_{j=1}^{n} a_{ij}b_{jl}\right)$$

$$= \sum_{j=1}^{n} k a_{ij}b_{jl}$$

The other side of the equation yields:

$$\left(kA\right)B = \sum_{j=1}^{n} \left(ka_{ij}\right)b_{jl}$$

$$= \sum_{j=1}^{n} k a_{ij}b_{jl}$$

It follows that $\mathrm{k}(AB) = (kA)B$. $\qquad\qquad\qquad\qquad\qquad\qquad\qquad\square$

Definition 4.6 The **identity matrix** $I_n \in \mathscr{M}_{n \times n}(\mathbb{F})$ is given by [4–7]:

$$I_n = (\delta_{ij})_{i,j} = \begin{pmatrix} 1 & 0 & \cdots & 0 & 0 & 0 & \cdots & 0 \\ 0 & 1 & \cdots & 0 & 0 & 0 & \cdots & 0 \\ \vdots & \vdots & \ddots & \vdots & \vdots & \vdots & \ddots & \vdots \\ 0 & 0 & \cdots & 1 & 0 & 0 & \cdots & 0 \\ 0 & 0 & \cdots & 0 & 1 & 0 & \cdots & 0 \\ 0 & 0 & \cdots & 0 & 0 & 1 & \cdots & 0 \\ \vdots & \vdots & \ddots & \vdots & \vdots & \vdots & \ddots & \vdots \\ 0 & 0 & \cdots & 0 & 0 & 0 & \cdots & 1 \end{pmatrix} \qquad (4.1)$$

with δ_{ij} the **Kronecker delta**:

$$\delta_{ij} = \begin{cases} 1, & \text{if } i = j \\ 0, & \text{if } i \neq j \end{cases}$$

Proposition 4.2 If $A \in \mathscr{M}_{m \times n}(\mathbb{F})$, $B \in \mathscr{M}_{n \times p}(\mathbb{F})$, then

$$AI_n = A, \quad I_n B = B$$

Proof Let c_{ij} be an element of AI_n with:

$$c_{ij} = a_{i1}\delta_{1j} + a_{i2}\delta_{2j} + \ldots + a_{ij}\delta_{jj} + \ldots + a_{in}\delta_{nj}$$

According to the definition of I_n $c_{ij} = a_{ij}$ proving that $AI_n = A$. $\square$

Definition 4.7 A matrix $A \in \mathcal{M}_{n \times n}(\mathbb{F})$ is **invertible** [10, 11] if there exist a matrix $B \in \mathcal{M}_{n \times n}(\mathbb{F})$ such that

$$AB = BA = I_n$$

In the above definition, if the matrix B exists, then this matrix is unique. This matrix is called the **inverse** of A and is denoted by A^{-1}.

Proposition 4.3 *If $A \in \mathcal{M}_{n \times n}(\mathbb{F})$ is an invertible matrix, then A^{-1} is invertible and $(A^{-1})^{-1} = A$.*

Proof The expression implies that A is invertible so $A^{-1}A = I_n$ and $AA^{-1} = I_n$ it follows that A^{-1} is also invertible whose inverse is A. $\square$

Proposition 4.4 *If $A, B \in \mathcal{M}_{n \times n}(\mathbb{F})$ are invertible matrices, then AB is invertible and*

$$(AB)^{-1} = B^{-1}A^{-1}$$

Proof The expression implies that A and B are invertible:

$$AA^{-1} = A^{-1}A = I_n$$
$$BB^{-1} = B^{-1}B = I_n$$

With:

$$(AB)\left(B^{-1}A^{-1}\right) = \left(A\left(BB^{-1}\right)\right)A^{-1}$$
$$= (AI_n)A^{-1}$$
$$= AA^{-1}$$
$$= I_n$$

And:

$$\left(B^{-1}A^{-1}\right)(AB) = \left(B^{-1}\left(A^{-1}A\right)B\right)$$
$$= \left(B^{-1}I_n\right)B$$
$$= B^{-1}B$$
$$= I_n$$

It is possible to conclude that $(AB)^{-1} = B^{-1}A^{-1}$. $\square$

Proposition 4.5 *Let $A \in \mathcal{M}_{n \times n}(\mathbb{F})$ is an invertible matrix and $c \in \mathbb{F}, c \neq 0$. Then cA is invertible and*

$$(cA)^{-1} = c^{-1}A^{-1}$$

Proof Since A is invertible

$$\left(c^{-1}A^{-1}\right)cA = c^{-1}A^{-1}cA$$
$$= cc^{-1}A^{-1}A$$
$$= 1I_n$$
$$= I_n$$

The other side of the equality is:

$$(cA)^{-1}(cA) = I_n$$

Therefore $(cA)^{-1} = c^{-1}A$. $\qquad\qquad\qquad\qquad\qquad\qquad\qquad\qquad\square$

Definition 4.8 If $A \in \mathcal{M}_{m \times n}(\mathbb{F})$, the **transpose** of A, denoted by A^{T} is the matrix $n \times m$ defined by

$$A^{\mathsf{T}} = \left(a_{ji}\right)_{i,j}$$

Proposition 4.6 *Let A, B matrices and $c \in \mathbb{F}$, when the next operations are well defined, we have:*

1. $(A + B)^{\mathsf{T}} = A^{\mathsf{T}} + B^{\mathsf{T}}$
2. $(cA)^{\mathsf{T}} = cA^{\mathsf{T}}$
3. $(AB)^{\mathsf{T}} = B^{\mathsf{T}}A^{\mathsf{T}}$
4. *If A is invertible, then A^{T} is invertible and $(A^{\mathsf{T}})^{-1} = \left(A^{-1}\right)^{\mathsf{T}}$*

Proof 1. Let $A = \left[a_{ij}\right] \in \mathcal{M}_{m \times n}$, $B = \left[b_{ij}\right] \in \mathcal{M}_{m \times n}$, $A^T = \left[a_{ji}\right] \in \mathcal{M}_{n \times m}$ and $B^T = \left[b_{ji}\right] \in \mathcal{M}_{n \times m}$, then:

$$(A + B)^T = \left(a_{ij} + b_{ij}\right)^T$$
$$= a_{ji} + b_{ji}$$

And

$$A^T + B^T = a_{ji} + b_{ji}$$

Allows to conclude that $(A + B)^T = A^T + B^T$

2. Consider $A = \left[a_{ij}\right] \in \mathcal{M}_{m \times n}$, $A^T = \left[a_{ji}\right] \in \mathcal{M}_{n \times m}$ and $c = c^T \in \mathbb{F}$ making the operations $(cA)^T = c^T A^T = cA^T$ proves that $(cA)^T = cA^T$
3. Let $A = \left[a_{ij}\right] \in \mathcal{M}_{m \times n}$, $B = \left[b_{jk}\right] \in \mathcal{M}_{n \times p}$, Defining

$$C = (AB)^T$$

$$c_{ji} = \sum_{k=1}^{n} a_{ik}b_{kj}$$

And

$$D = B^T A^T$$

$$d_{ji} = \sum_{k=1}^{n} b_{kj} a_{ik}$$

Shows that $\sum_{k=1}^{n} a_{ik} b_{kj} = \sum_{k=1}^{n} b_{kj} a_{ik}$ so $(AB)^T = B^T A^T$

4. Consider $A = \left[a_{ij} \right] \in \mathscr{M}_{n \times n}$ is invertible, it follows:

$$A^T \left(A^{-1} \right)^T = \left(A^{-1} A \right)^T$$
$$= I_n^T$$
$$= I$$

Then A^T has the inverse $\left(A^T \right)^{-1} = \left(A^{-1} \right)^T$. $\qquad\qquad\qquad\square$

Definition 4.9 Let V a finite-dimensional vector space over a field $\mathbb{F}$, with dim $V = n$. Let $\{v_1, \ldots, v_n\}$ and $\{v_1', \ldots, v_n'\}$ basis of V. The **change of basis matrix**.

4.4 Gauss-Jordan Method

Definition 4.10 Let $A = \left(a_{ij} \right)_{i,j} \in \mathscr{M}_{m \times n}(\mathbb{F})$ and $c \in \mathbb{F}$. An **elementary row operation** is function e which associated with each $A \in \mathscr{M}_{m \times n}(\mathbb{F})$ an $e(A) \in \mathscr{M}_{m \times n}(\mathbb{F})$, i.e.

1. Multiplication of one row of A by a non-zero scalar c: $e(A)_{ij} = A_{ij}$, if $i \neq r$, $e(A)_{rj} = c A_{rj}$.
2. Replacement of the $r - th$ row of A by row r plus c times row s. c is any scalar and $r \neq s$: $e(A)_{ij} = A_{ij}$, if $i \neq r$, $e(A)_{rj} = A_{rj} + c A_{sj}$.
3. Interchange of two rows of A: $e(A)_{ij} = A_{ij}$, if $i \neq r, i \neq s$, $e(A)_{rj} = A_{sj}$, $e(A)_{sj} = A_{rj}$.

Definition 4.11 It said that an **elementary matrix** is a matrix which is obtained after operating the identity matrix with one of the three elementary row operations. There are three types of elementary matrices, which correspond to three types of row operations:

1. **Row switching**: A row i within the matrix I_n can be switched with another row j (**elemental matrix of first type**).

$$I_{ij} = \begin{pmatrix} 1 & 0 & \cdots & 0 & 0 & 0 & \cdots & 0 \\ 0 & 1 & \cdots & 0 & 0 & 0 & \cdots & 0 \\ \vdots & \vdots & \ddots & \vdots & \vdots & \vdots & \ddots & \vdots \\ 0 & 0 & \cdots & 0 & 1 & 0 & \cdots & 0 \\ 0 & 0 & \cdots & 1 & 0 & 0 & \cdots & 0 \\ 0 & 0 & \cdots & 0 & 0 & 1 & \cdots & 0 \\ \vdots & \vdots & \ddots & \vdots & \vdots & \vdots & \ddots & \vdots \\ 0 & 0 & \cdots & 0 & 0 & 0 & \cdots & 1 \end{pmatrix} \tag{4.2}$$

2. **Row multiplication**: Each element in a row i of I_n can be multiplied by a non-zero constant $c \in \mathbb{F}$ (**elemental matrix of second type**).

$$I_i(c) = \begin{pmatrix} 1 & 0 & \cdots & 0 & 0 & 0 & \cdots & 0 \\ 0 & 1 & \cdots & 0 & 0 & 0 & \cdots & 0 \\ \vdots & \vdots & \ddots & \vdots & \vdots & \vdots & \ddots & \vdots \\ 0 & 0 & \cdots & 1 & 0 & 0 & \cdots & 0 \\ 0 & 0 & \cdots & 0 & c & 0 & \cdots & 0 \\ 0 & 0 & \cdots & 0 & 0 & 1 & \cdots & 0 \\ \vdots & \vdots & \ddots & \vdots & \vdots & \vdots & \ddots & \vdots \\ 0 & 0 & \cdots & 0 & 0 & 0 & \cdots & 1 \end{pmatrix} \tag{4.3}$$

3. **Row addition**: A row i of I_n can be replaced by the sum of that row i and a multiple of another row j (**elemental matrix of third type**).

$$I_{ij}(c) = \begin{pmatrix} 1 & 0 & \cdots & 0 & 0 & 0 & \cdots & 0 \\ 0 & 1 & \cdots & 0 & 0 & 0 & \cdots & 0 \\ \vdots & \vdots & \ddots & \vdots & \vdots & \vdots & \ddots & \vdots \\ 0 & 0 & \cdots & 1 & c & 0 & \cdots & 0 \\ 0 & 0 & \cdots & 0 & 1 & 0 & \cdots & 0 \\ 0 & 0 & \cdots & 0 & 0 & 1 & \cdots & 0 \\ \vdots & \vdots & \ddots & \vdots & \vdots & \vdots & \ddots & \vdots \\ 0 & 0 & \cdots & 0 & 0 & 0 & \cdots & 1 \end{pmatrix} \tag{4.4}$$

Lemma 4.1 *The elementary matrices are invertible, i.e.:*

1. $\left(I_{ij}\right)^{-1} = I_{ij}$
2. $(I_i(c))^{-1} = I_i(c^{-1})$
3. $\left(I_{ij}(c)\right)^{-1} = I_{ij}(-c)$

Proof 1. Consider the matrix I_{ij} defined by the row swapping $r_i = r_j$, then the next expression is true:

$$I_{ij} I_{ij} = I$$

2. Consider the elementary matrix $I_i(c)$ formed by the row operation $r_i = cr_i$ and the matrix $I_i^{-1}(c)$ defined by the row operation $r_i = \frac{1}{c}r_i$, then:

$$I_i(c) I_i^{-1}(c) = I_i^{-1}(c) I_i(c)$$
$$= I$$

3. The elementary matrix $I_{ij}(c)$ is defined by the row operation $r_i = r_i + ar_j$ and another matrix $I_{ij}^{-1}(c)$ defined by $r_i = r_i - ar_j$, so that:

$$I_{ij}(c) I_{ij}^{-1}(c) = I_{ij}^{-1}(c) I_{ij}(c)$$
$$= I \qquad \qquad \square$$

Proposition 4.7 *Let $A = (a_{ij})_{i,j} \in \mathcal{M}_{m \times n}(\mathbb{F})$ and X an elemental matrix. The product XA is equivalent to realize an elemental row operation to matrix A.*

Proof Let r be an elementary row operation and $X = r(I)$ the resulting elementary matrix of such operation. The proof will consider the three previously stated elementary matrix operations, in the first case if the rows $\alpha \neq \beta$ in the operation $\alpha = \beta$ make the next elementary matrix:

$$X_{ij} = \begin{cases} \delta_{ik} & i \neq \alpha \\ \delta_{\beta k} & i = \alpha \end{cases}$$

Thus

$$XA = \sum_{k=1}^{m} X_{ik} A_{kj} = \begin{cases} A_{ik} & i \neq \alpha \\ A_{\beta j} & i = \alpha \end{cases}$$

The second case has the operation $\alpha = c\beta$ with $\alpha \neq \beta$ having the elementary matrix be:

$$X_{ij} = \begin{cases} \delta_{ik} & i \neq \alpha \\ c\delta_{\beta k} & i = \alpha \end{cases}$$

And the operation

$$XA = \sum_{k=1}^{m} X_{ik} A_{kj} = \begin{cases} A_{ik} & i \neq \alpha \\ c A_{\beta j} & i = \alpha \end{cases}$$

For the final case consider the rows $\alpha \neq \beta$ in the operation $\alpha = \alpha + c\beta$, then the elementary matrix is:

$$X_{ik} = \begin{cases} \delta_{ik} & i \neq \alpha \\ \delta_{\alpha k} + c\delta_{\beta k} & i = \alpha \end{cases}$$

Making

$$XA = \sum_{k=1}^{m} X_{ik} A_{kj} = \begin{cases} A_{ik} & i \neq \alpha \\ A_{\alpha j} + c A_{\beta j} & i = \alpha \end{cases}$$

showing that $XA = r(A)$.   □

Definition 4.12 Let $A = (a_{ij})_{i,j} \in \mathcal{M}_{m \times n}(\mathbb{F})$. The matrix A is called **row-reduced** if:

1. The first non-zero entry in each non-zero row of A is equal to 1.
2. Each column of A which contains the leading non-zero entry of some row has all its other entries 0.

The **Gauss-Jordan elimination method** consists in to apply successively elementary row operations in order to bring the matrix A to a row-reduced matrix.

Finally, we consider the algorithm to determine if an matrix is invertible and if it is possible, to calculate **the inverse of a square matrix**. Let $A = (a_{ij})_{i,j} \in \mathcal{M}_{n \times n}(\mathbb{F})$. We place the identity matrix I_n to the right of the matrix A and we obtain a matrix $n \times 2n$. This matrix (augmented matrix) is defined by

$$M_A = [A \mid I_n]$$

Now, it is necessary to realize elementary row operations to the augmented matrix M_A in a consecutive way until to obtain a row-reduced matrix. If A is invertible, this last obtained matrix is equal to $I_n = A^{-1}$. Therefore, if in the first block of the last matrix, after the process of elementary operations, to appear the identity matrix, then the matrix A is invertible and A^{-1} is the matrix in the second block. If in the first block, the matrix is not the identity, then the matrix A is not an invertible matrix [10, 11].

4.5 Definitions

Definition 4.13 A **linear equation** over $\mathbb{F}$ is an expression of the form:

$$a_1 x_1 + a_2 x_2 + \ldots + a_n x_n = b \tag{4.5}$$

with $a_1, a_2, \ldots, a_n, b \in \mathbb{F}$ and $x_1, x_2, \ldots, x_n$ are independent variables. A **solution** of (4.5) is $(\alpha_1, \alpha_1, \ldots, \alpha_n)$ such that:

$$a_1 \alpha_1 + a_2 \alpha_2 + \ldots + a_n \alpha_n = b$$

Now, we consider a **system of m linear equations with n unknowns**:

$$
\begin{aligned}
a_{11} x_1 + a_{12} x_2 + \ldots + a_{1n} x_n &= b_1 \\
a_{21} x_1 + a_{22} x_2 + \ldots + a_{2n} x_n &= b_2 \\
&\vdots \\
a_{m1} x_1 + a_{m2} x_2 + \ldots + a_{mn} x_n &= b_m
\end{aligned}
\tag{4.6}
$$

where $a_{ij} \in \mathbb{F}$, $1 \le i \le m, 1 \le j \le n$ and $x_1, x_2, \ldots, x_n$ are independent variables. If $b_1 = b_2 = \ldots = b_m = 0$ the system (4.6) is said to be **homogeneous**. A **solution** of (4.6) is $(\alpha_1, \alpha_1, \ldots, \alpha_n)$ such that:

$$
\begin{aligned}
a_{11} \alpha_1 + a_{12} \alpha_2 + \ldots + a_{1n} \alpha_n &= b_1 \\
a_{21} \alpha_1 + a_{22} \alpha_2 + \ldots + a_{2n} \alpha_n &= b_2 \\
&\vdots \\
a_{m1} \alpha_1 + a_{m2} \alpha_2 + \ldots + a_{mn} \alpha_n &= b_m
\end{aligned}
$$

Finally, **the solution** or **general solution** of (4.6) consists in all the possible solutions. If the system of linear equations is homogeneous, then we have two possibilities:

1. The unique solution is the **trivial solution**.
2. There are more solutions.

If the system of linear equations is not homogeneous, there are two possibilities:

1. The system is inconsistent (there are not solutions).
2. The system is consistent (there are solutions). In this case, there are two situations: the solution is unique or there exists more solutions.

Theorem 4.1 *Let a homogeneous system of m linear equations with n unknowns. If $n > m$ then the system has only one (nontrivial) solution in $\mathbb{F}$.*

Proof The proof is left to the reader as an exercise. $\square$

References

1. Hoffman, K., Kunze, R.: Linear Algebra. Prentice-Hall, Englewood Cliffs (1971)
2. Axler, S.: Linear Algebra Done-Right. Springer, New York (1997)
3. Grossman, S.: Elementary Linear Algebra, Wadsworth Publishing, Belmont (1987)
4. Lipschutz, S.: Linear Algebra. McGraw Hill, New York (2009)
5. Nef, W.: Linear Algebra. Dover Publications, New York (1967)
6. Lang, S.: Linear Algebra. Springer, New York (2004)
7. Anton, H.: Elementary Linear Algebra. Wiley, New York (2010)
8. Whitelaw, T.: Introduction to Linear Algebra. Chapman & Hall/CRC, Boca Raton (1992)
9. Kolman, B.: Introductory Linear Algebra an Applied First Course. Pearson Education (2008)
10. Gantmacher, F.R.: The Theory of Matrices, vol. 1. Chelsea, New York (1959)
11. Gantmacher, F.R.: The Theory of Matrices, vol. 2. Chelsea, New York (1964)
12. Nering, E.: Linear Algebra and Matrix Theory. Wiley, New York (1970)

Chapter 5
Permutations and Determinants

Abstract This chapter introduces basic concepts about permutations and determinants of matrices, the first part contains permutation groups explained through various definitions and theorems, the second part is about determinants of matrices, a method to obtain the determinant of a matrix is given along various other useful results.

5.1 Permutations Group

Definition 5.1 Let X a nonempty set. The **permutations group** in X, denoted by S_X is the set of bijective functions of X itself. The elements of S_X are called **permutations** (of the elements of X) [1–4].

Proposition 5.1 *Let X a nonempty set.*

1. *$\sigma, \tau \in S_X$ then $\sigma \circ \tau \in S_X$.*
2. *$\forall \sigma, \tau, \rho \in S_X, (\sigma \circ \tau) \circ \rho = \sigma \circ (\tau \circ \rho)$.*
3. *There exists $id_X \in S_X$ such that $\sigma \circ id_X = id_X \circ \sigma = \sigma \ \forall \sigma \in S_X$.*
4. *Let $\sigma \in S_X$ there exists $\tau = \sigma^{-1} \in S_X$ such that $\sigma \circ \tau = \tau \circ \sigma = id_X$.*

Proof 1. The proof 1, 2 and 4 are left as an exercise to the reader.
2.
3. Having $\sigma \in S_X$ and $x \in X$, then $(\sigma \circ id_X)(x) = \sigma(id_X(x)) = \sigma(x)$ and $(id_X \circ \sigma)(x) = id_X(\sigma(x)) = \sigma(x)$ then it is possible to conclude $\sigma \circ id_X = id_X \circ \sigma = \sigma$
4. $\qquad\qquad\qquad\qquad\qquad\qquad\qquad\qquad\qquad\qquad\qquad\qquad\qquad\qquad \square$

Corollary 5.1 *Let $\sigma, \tau, \rho \in S_X$.*

1. *If $\sigma \circ \tau = \sigma \circ \rho$, then $\tau = \rho$.*
2. *If $\sigma \circ \rho = \tau \circ \rho$, then $\sigma = \tau$.*

© Springer Nature Switzerland AG 2019
R. Martínez-Guerra et al., *Algebraic and Differential Methods for Nonlinear Control Theory*, Mathematical and Analytical Techniques with Applications to Engineering, https://doi.org/10.1007/978-3-030-12025-2_5

Proof 1. $\sigma \circ \tau = \sigma \circ \rho \Rightarrow \sigma^{-1} \circ (\sigma \circ \tau) = \sigma^{-1} \circ (\sigma \circ \rho) \Rightarrow \left(\sigma^{-1} \circ \sigma\right) \circ \tau = \left(\sigma^{-1} \circ \sigma\right) \circ \rho \Rightarrow id_X \circ \tau = id_X \circ \rho \Rightarrow \tau = \rho$

2. The procedure is similar to 1 and it is left as an exercise to the reader. □

Proposition 5.2 *If X is a finite set with n elements, then S_X has $n! = 1 \cdot 2 \cdot \ldots \cdot n$ elements. In this case $S_X = S_n$.*

Proof It is left as an exercise to the reader. □

Remark 5.1 Let suppose $n > 1$. If $X = \{1, 2, \ldots, n\}$ and $\varphi \in S_n$, then φ is a bijection of X in itself:

$$\varphi : X \longrightarrow X$$
$$1 \longmapsto i_1$$
$$2 \longmapsto i_2$$
$$\vdots$$
$$n \longmapsto i_n$$

We write:

$$\varphi = \begin{pmatrix} 1 & 2 & \cdots & n \\ i_1 & i_2 & \cdots & i_n \end{pmatrix}$$

Let $n, k \in \mathbb{N}$, with $k \le n$. Let $i_1, i_2, \ldots, i_k$ different integers in $X = \{1, 2, \ldots, n\}$. We write

$$\left(i_1 \, i_2 \, \cdots \, i_k\right)$$

to represent the permutation $\sigma \in S_n$ given by

$$\sigma(i_1) = i_2, \sigma(i_2) = i_3, \ldots, \sigma(i_{k-1}) = i_k, \sigma(i_k) = i_1$$

and $\sigma(x) = x \quad \forall x \in X, x \notin \{i_1, i_2, \ldots, i_k\}$.

Definition 5.2 It said that a permutation $\left(i_1 \, i_2 \, \cdots \, i_k\right)$ is called **cycle of order** k or k-**cycle**. If $k = 2$, the 2-cycle is called **transposition**.

Proposition 5.3 *If σ is a k-cycle then*

$$\sigma^k = \sigma \circ \cdots \circ \sigma = id$$

Proof It is left to the reader as an exercise. □

Definition 5.3 It is said that two cycles are **disjoint** if they have elements in common.

Lemma 5.1 *Let $\sigma, \tau \in S_n$. If σ and τ are disjoint cycles then $\sigma \circ \tau = \tau \circ \sigma$. In this case we say that σ and τ commute.*

Proof It is left to the reader as an exercise. □

Theorem 5.1 *Every permutation in S_n is a product of disjoint cycles pairs. This is uniquely except the order and 1-cycle [2, 5–7].*

Proof It is left to the reader as an exercise. □

Theorem 5.2 *Every permutation in S_n is a product of transpositions.*

Proof It is left to the reader as an exercise. □

Theorem 5.3 *Any permutation in S_n is the product of an even number of transpositions or product of an odd number of transpositions but not both.*

Proof Consider the permutations:

$$(ab)(ac_1 \cdots c_h bd_1 \cdots d_k) = (bd_1 \cdots d_k)(ac_1 \cdots c_h)$$

Making h or k be 0 causes that c or d does not occur, using in both sides $i \in \{1, 2, \ldots, n\}$ shows that the equation holds. Since $(ab)^{-1} = ab$ multiplying both sides of the equation by (ab) yields:

$$(ab)(bd_1 \cdots d_k)(ac_1 \cdots c_h) = (ac_1 \cdots c_h bd_1 \cdots d_k)$$

The value of N is $N(ac_1 \cdots c_h bd_1 \cdots d_k) = h + k + 1$ and $N((bd_1 \cdots d_k)(ac_1 \cdots c_h)) = h + k$, it follows that $N(ab)(\alpha) = N(\alpha) - 1$ if a and b occur in the same operation in the decomposition of α, otherwise $N(ab)(\alpha) = N(\alpha) + 1$ if a and b are present in different operations, hence if α is a product of m transpositions then, $N(1) = 0$, $N(\alpha) = \sum_{i=1}^{m} \varepsilon_i$ where $\varepsilon_i = \pm 1$, changing an $\varepsilon_i = -1$ for 1 equals adding 2 thus, it does not affect the parity, if the same is applied to $\varepsilon_i = 1$ the final sum will be m that has the same parity as $N(\alpha)$. □

Definition 5.4 The **parity** of a permutation $\sigma \in S_n$ is:

1. An **even permutation** if the product of an even number of transpositions.
2. An **odd permutation** if the product of an odd number of transpositions.

5.2 Determinants

Let $\sigma \in S_n$ the permutations group of $\{1, \ldots, n\}$. The **sign of** σ is given by the following rule:

$$(-1)^{\sigma} = \begin{cases} 1, & \text{if } \sigma \text{ is even.} \\ -1, & \text{if } \sigma \text{ is odd.} \end{cases}$$

Definition 5.5 Let $A = \left(a_{ij}\right)_{i,j} \in \mathscr{M}_{n \times n}(\mathbb{F})$. The **determinant** of A is:

$$\det(A) = \sum_{\sigma \in S_n} (-1)^{\sigma} a_{1\sigma(1)} a_{2\sigma(2)} \cdots a_{n\sigma(n)}$$

Remark 5.2 The determinat of A is denoted by $|A|$ or by [7–10]

$$\begin{vmatrix} a_{11} & a_{12} & \cdots & a_{1n} \\ a_{21} & a_{22} & \cdots & a_{2n} \\ \vdots & & & \vdots \\ a_{n1} & a_{n2} & \cdots & a_{nn} \end{vmatrix}$$

Definition 5.6 Let $A = \left(a_{ij}\right)_{i,j} \in \mathscr{M}_{n \times n}(\mathbb{F})$.

Theorem 5.4 *Let $A = \left(a_{ij}\right)_{i,j} \in \mathscr{M}_{n \times n}(\mathbb{F})$. The determinant of A is equal to the sum of the obtained products through the multiplication of elements of any row or column by their respectively cofactors.*

Proof Let $A = \left[a_{ij}\right]$ and A_{ij} denote the cofactor of a_{ij}, then for any i or j

$$|A| = a_{i1} A_{i1} + \cdots + a_{in} A_{in}$$
$$= a_{1j} A_{1j} + \cdots + a_{nj} A_{nj}$$

Having $|A| = |A^T|$ leads to

$$|A| = a_{i1} A_{i1}^* + \cdots + a_{in} A_{in}^*$$

where A_{ij}^* is the sum of terms involving no entry of the ith row of A. Consider $i = n$ and $j = n$ makes the sum of terms in $|A|$ containing a_{nn} is

$$a_{nn} A_{nn}^* = a_{nn} \sum_{\sigma} sign\,(\sigma)\, a_{1\sigma(1)} a_{2\sigma(2)} \cdots a_{n-1\sigma(n_1)}$$

The sum of all permutations $\sigma \in S_n$ for $\sigma(n) = n$ which equals to adding over the permutations $\{1, \ldots, n-1\}$ so $A_{nn}^* = |M_{nn}| = (-1)^{n+n} |M_{nn}|$. Considering every other i and j the ith row is exchanged with each succeeding row and the jth column is exchanged with each succeeding column, this operation does not change the determinant $|M_{ij}|$, but the sign of $|A|$ and A_{ij}^* does change by $n-1$ and $n-j$ accordingly, so

$$A_{ij}^* = (-1)^{n-i+n-j} |M_{ij}|$$
$$= (-1)^{i+j} |M_{ij}|$$

Showing that the theorem is true.   □

Proposition 5.4 Let $A = (a_{ij})_{i,j} \in \mathcal{M}_{n \times n}(\mathbb{F})$.

$$\det(A^\mathsf{T}) = \det(A)$$

Proof Let $A = [a_{ij}]$ and $A^T = [b_{ij}]$ with $a_{ji} = b_{ij}$, then,

$$\left|A^T\right| = \sum_{\sigma \in S_n} sign\,(\sigma)\,b_{1\sigma(1)} \cdots b_{n\sigma(n)} = \sum_{\sigma \in S_n} sign\,(\sigma)\,a_{\sigma(1)1} \cdots a_{\sigma(n)n}$$

With $\tau = \sigma^{-1}$ then $sign\,(\tau) = sign\,(\sigma)$ and $a_{\sigma(1)1} \cdots a_{\sigma(n)n} = a_{1\tau(1)} \cdots a_{n\tau(n)}$ then

$$\left|A^T\right| = \sum_{\sigma \in S_n} sign\,(\tau)\,a_{1\tau(1)} \cdots a_{n\tau(n)}$$

But σ spans across all $S_{n,\tau} = \sigma^{-1}$ that also spans through all elements of S_n thus $|A| = \left|A^T\right|$. □

Definition 5.7 Let $A = (a_{ij})_{i,j} \in \mathcal{M}_{n \times n}(\mathbb{F})$.

1. A is said to be **superior triangular** if $a_{ij} = 0$ and $i > j$.
2. A is said to be **inferior triangular** if $a_{ij} = 0$ and $i < j$.
3. A is said to be **triangular** if it is superior triangular or inferior triangular.
4. A is said to be **diagonal** if $a_{ij} = 0$ and $i \neq j$.

Proposition 5.5 *The determinant of a triangular matrix is the product of its entries in the diagonal [1, 10–12].*

Proof It is left to the reader as an exercise. □

Corollary 5.2 *The determinant of a diagonal matrix is the product of its entries in the diagonal.*

Proof It is left to the reader as an exercise. □

Proposition 5.6 Let $A = (a_{ij})_{i,j} \in \mathcal{M}_{n \times n}(\mathbb{F})$ [5, 9, 12].

1. *If B is obtained through the interchange between two rows (or columns) of A, then $\det B = - \det A$.*
2. *If B is obtained through the multiplication between a row (or a column) of A and a scalar c, then $\det B = c \det A$.*
3. *If B is obtained through the sum of a multiple of row (or a column) of A and then $\det B = \det A$.*

Proof 1. Let τ be a transposition that interchanges the numbers corresponding to the columns of A that are exchanged, if $A = \begin{bmatrix} a_{ij} \end{bmatrix}$ and $B = \begin{bmatrix} b_{ij} \end{bmatrix}$ then $b_{ij} = a_{i\tau(j)}$, for a permutation σ

$$b_{1\sigma(1)} \cdots b_{n\sigma(n)} = a_{1(\tau\circ\sigma)(1)} \cdots a_{n(\tau\circ\sigma)(n)}$$

So

$$|B| = \sum_{\sigma\in S_n} sign\,(\sigma)\, b_{1\sigma(1)} \cdots b_{n\sigma(n)} = \sum_{\sigma\in S_n} sign\,(\sigma)\, a_{1(\tau\circ\sigma)(1)} \cdots a_{n(\tau\circ\sigma)(n)} \ \ .$$

Being τ an odd permutation makes $sign\,(\tau\circ\sigma) = sign\,(\tau)\,sign\,(\sigma) = -sign\,(\sigma)$ and $sign\,(\sigma) = -sign\,(\tau\circ\sigma)$, hence

$$|B| = \sum_{\sigma\in S_n} sign\,(\tau\circ\sigma)\, a_{1(\tau\circ\sigma)(1)} \cdots a_{n(\tau\circ\sigma)(n)}$$

Since σ spans through S_n, $\tau\circ\sigma$ which also spans across S_n, thus $|B| = -|A|$

2. Multiplying the jth row by k, then every element of $|A|$ is multiplied by k so $|B| = k\,|A|$ so

$$|B| = \sum_{\sigma} sign\,(\sigma)\, a_{1i_1} \cdots ka_{ji_n} \cdots a_{ni_n}$$

$$= k\sum_{\sigma} sign\,(\sigma)\, a_{1i_1} \cdots a_{ni_n}$$

$$= k\,|A|$$

3. The k-th row is added c times to the j-th row of A, using ^ to denote the j-th position of a term in the determinant makes:

$$|B| = \sum_{\sigma} sign\,(\sigma)\, a_{1i_1} \cdots c\hat{a}_{ki_k} + a_{ji_j} \cdots a_{ni_n}$$

$$= c\sum_{\sigma} sign\,(\sigma)\, a_{1i_1} \cdots \hat{a}_{ki_k} \cdots a_{ni_n} + \sum_{\sigma} sign\,(\sigma)\, a_{1i_1} \cdots a_{ji_j} \cdots a_{ni_n}$$

With this result it is possible to conclude that $|B| = c\cdot 0 + |A| = |A|$. $\square$

Theorem 5.5 *Let*

$$det : \mathcal{M}_{n\times n}(\mathbb{F}) \longrightarrow \mathbb{F}$$

with $A \longmapsto det\,A$. *Then*

$$det\,(AB) = det\,(A)\,det\,(B)$$

Proof If A is singular then AB is also singular, so $|AB| = 0 = |A| |B|$, if A is nonsingular $A = E_n \cdots E_1$ then

$$|AB| = |E_n \cdots E_1 B| = |E_n| \cdots |E_1| |B| = |A| |B| \qquad \square$$

Corollary 5.3 *Let* $A \in \mathcal{M}_{n \times n}(\mathbb{F})$. *If A is an invertible matrix then* $\det A \neq 0$ *and*

$$\det \left(A^{-1} \right) = (\det A)^{-1}$$

Proof Since A is invertible $AA^{-1} = I_n$, then $|A| \left| A^{-1} \right| = |I_n| = 1$ so $|A| \neq 0$ and $\left| A^{-1} \right| = |A|^{-1}$. $\qquad \square$

Corollary 5.4 *Let* $A, P \in \mathcal{M}_{n \times n}(\mathbb{F})$. *If P is an invertible matrix then*

$$\det \left(P^{-1} A P \right) = \det A$$

Proof The previous statement leads to:

$$\begin{aligned} \left| P^{-1} A P \right| &= \left| P^{-1} \right| |A| |P| \\ &= |P|^{-1} |A| |P| \\ &= |P|^{-1} |P| |A| \\ &= |I_n| |A| \\ &= 1 |A| \\ &= |A| \end{aligned}$$

Showing that the corollary is true. $\qquad \square$

Remark 5.3 Note that if $T : V \longrightarrow V$ is a linear operator, where V is a finite-dimensional vector space in $\mathbb{F}$ then, it is possible to define the determinant of T:

$$\det T = \det A$$

where A is the associated matrix to T respect any basis of V.

Definition 5.8 Let $A = \left(a_{ij} \right)_{i,j} \in \mathcal{M}_{n \times n}(\mathbb{F})$. The **adjoint matrix** of A is the transposed matrix of the matrix cofactors, i.e.:

$$\mathrm{Adj}\,(A) = \begin{pmatrix} A_{11} & A_{21} & \cdots & A_{n1} \\ A_{21} & A_{22} & \cdots & A_{n2} \\ \vdots & \vdots & \ddots & \vdots \\ A_{1n} & A_{2n} & \cdots & A_{nm} \end{pmatrix}$$

where A_{ij} is the cofactor of a_{ij}.

Theorem 5.6 *Let $A \in \mathcal{M}_{n \times n}(\mathbb{F})$, then*

$$A \cdot Adj(A) = \det(A) \cdot I_n$$

Proof Having $A = [a_{ij}]$ and $A\text{adj}(A) = [b_{ij}]$ makes the ith row of A be $(a_{i1}, \ldots, a_{in})$ and the jth column of adj (A) is $(A_{j1}, \ldots, A_{jn})$, the b_{ij} element of $A\text{adj}(A)$ is given by $b_{ij} = a_{i1}A_{j1} + a_{i2}A_{j2} + \cdots + a_{in}A_{jn}$ which can also be represented as

$$b_{ij} = \begin{cases} |A| & if\ i = j \\ 0 & if\ i \neq j \end{cases}$$

The matrix $A\text{adj}(A)$ is a diagonal matrix with every diagonal element of $|A|$ showing that $A(\text{adj}(A)) = |A|\,I$ and $(\text{adj}(A)\,A) = |A|\,I$.   □

Corollary 5.5 *Let $A \in \mathcal{M}_{n \times n}(\mathbb{F})$, then A is invertible if and only if $\det(A) \neq 0$ and*

$$A^{-1} = \frac{1}{\det(A)} \cdot Adj(A)$$

Proof It is left to the reader as an exercise.   □

Finally, we consider a system of linear equations with n equations and n variables:

$$\left.\begin{array}{c} a_{11}x_1 + a_{12}x_2 + \cdots + a_{1n}x_n = b_1 \\ a_{21}x_1 + a_{22}x_2 + \cdots + a_{2n}x_n = b_2 \\ \vdots \\ a_{n1}x_1 + a_{n2}x_2 + \cdots + a_{nn}x_n = b_n \end{array}\right\} \tag{5.1}$$

Let

$$A = (a_{ij})_{i,j} = \begin{pmatrix} a_{11} & a_{21} & \cdots & a_{1n} \\ a_{21} & a_{22} & \cdots & a_{1n} \\ \vdots & \vdots & \ddots & \vdots \\ a_{1n} & a_{2n} & \cdots & a_{nn} \end{pmatrix}$$

where A is called the **coefficients matrix** of the system (5.1). Let $\Delta = \det(A)$ for $i \in \{1, \ldots, n\}$:

$$\Delta_i = \begin{vmatrix} a_{11} & \cdots & a_{1i-1} & b_1 & a_{1i+1} & \cdots & a_{1n} \\ a_{21} & \cdots & a_{2i-1} & b_2 & a_{2i+1} & \cdots & a_{2n} \\ \vdots & \ddots & \vdots & \vdots & \vdots & \ddots & \vdots \\ a_{n1} & \cdots & a_{ni-1} & b_n & a_{ni+1} & \cdots & a_{nn} \end{vmatrix}$$

In linear algebra, **Cramer's rule** is an explicit formula for the solution of a system of linear equations with as many equations as unknowns, valid whenever the system has a unique solution. It expresses the solution in terms of the determinants of the (square) coefficient matrix and of matrices obtained from it by replacing one column by the vector of right hand sides of the equations. The following theorem explains the general method for Cramer's rule.

Theorem 5.7 *If $\Delta \neq 0$, then the system of linear equations* (5.1) *has a unique solution given by:*

$$x_i = \frac{\Delta_i}{\Delta} \quad , \quad 1 \leq i \leq n$$

Proof The system $AX = B$ has a unique solution if and only if A is invertible which in turn is only invertible if $\Delta = |A| \neq 0$, suppose that $\Delta \neq 0$, this causes $A^{-1} = (1/\Delta) \operatorname{adj}(A)$. Multiplying $AX = B$ by A^{-1} yields:

$$X = A^{-1}AX$$
$$= (1/\Delta) \operatorname{adj}(A) B$$

The ith row of $(1/\Delta) \operatorname{adj}(A) B$ is $(1/\Delta)(A_{1i}, \ldots, A_{ni})$, if $B = (b_1, \ldots, b_n)^T$, then:

$$x_i = (1/\Delta)(b_1 A_{1i} + \cdots + b_n A_{ni})$$

And since $\Delta_i = b_1 A_{1i} + \cdots + b_n A_{ni}$ yields $x_i = \frac{\Delta_i}{\Delta}$. $\qquad\square$

References

1. Hoffman, K., Kunze, R.: Linear Algebra. Prentice-Hall (1971)
2. Axler, S.: Linear Algebra Done-Right. Springer (1997)
3. Nef, W.: Linear Algebra. Dover Publications (1967)
4. kolman, B.: Introductory Linear Algebra an Applied First Course. Pearson Education (2008)
5. Lipschutz, S.: Linear Algebra. McGraw Hill (2009)
6. Gantmacher, F.R.: The Theory of Matrices, vol. 2. Chelsea, New York (1964)
7. Halmos, P.R.: Finite-Dimensional Vector Spaces. Courier Dover Publications (2017)
8. Whitelaw, T.: Introduction to Linear Algebra. Chapman & Hall/CRC (1992)
9. Nering, E.: Linear Algebra and Matrix Theory. Wiley (1970)
10. Anton, H.: Elementary Linear Algebra. Wiley (2010)
11. Grossman, S.: Elementary Linear Algebra. Wadsworth Publishing (1987)
12. Gantmacher, F.R.: The Theory of Matrices, vol. 1. Chelsea, New York (1959)

Chapter 6
Vector and Euclidean Spaces

Abstract This chapter is about vector spaces and Euclidean spaces, the first part presents definitions and concepts about vector spaces and subspaces, followed by generated subspace, linear dependence and independence, bases and dimension, quotient space, Euclidean spaces and finally the GramSchidt process is given, every section includes several proposed and solved exercises.

6.1 Vector Spaces and Subspaces

Definition 6.1 It said that a field [1–5] is a set $\mathbb{F}$ with two operations:

$$+ : \mathbb{F} \times \mathbb{F} \to \mathbb{F}$$

$$\cdot : \mathbb{F} \times \mathbb{F} \to \mathbb{F}$$

such that:

1. Addition is **commutative**:

$$x + y = y + x \quad \forall x, y \in \mathbb{F}$$

2. Addition is **associative**:

$$x + (y + z) = (x + y) + z \quad \forall x, y, z \in \mathbb{F}$$

3. There is a unique element $0 \in \mathbb{F}$ called **identity element** (or **neutral element**) for addition, such that:

$$x + 0 = 0 + x = x \quad \forall x \in \mathbb{F}$$

© Springer Nature Switzerland AG 2019
R. Martínez-Guerra et al., *Algebraic and Differential Methods for Nonlinear Control Theory*, Mathematical and Analytical Techniques with Applications to Engineering, https://doi.org/10.1007/978-3-030-12025-2_6

4. For all $x \in \mathbb{F}$ there exists a unique element $(-x) \in \mathbb{F}$ such that:

$$x + (-x) = (-x) + x = 0$$

5. Multiplication is **commutative**:

$$x \cdot y = y \cdot x \quad \forall x, y \in \mathbb{F}$$

6. Multiplication is **associative**:

$$x \cdot (y \cdot z) = (x \cdot y) \cdot z \quad \forall x, y, z \in \mathbb{F}$$

7. There exists a unique non-zero element $1 \in \mathbb{F}$ called **identity element** (or **neutral element**) for multiplication, such that:

$$x \cdot 1 = 1 \cdot x = x \quad \forall x \in \mathbb{F}$$

8. To each non-zero $x \in \mathbb{F}$ there corresponds a unique element x^{-1} (equivalently $1/x$) such that:
$$xx^{-1} = x^{-1}x = 1$$

9. Multiplication **distributes** over addition:

$$x \cdot (y + z) = x \cdot y + x \cdot z \quad \forall x, y, z \in \mathbb{F}$$

Remark 6.1 Note that in $\mathbb{C}$, $0 = 0 + 0i$ and $1 = 1 + 0i$.

Example 6.1 $\mathbb{Q}$, $\mathbb{R}$ and $\mathbb{C}$ are examples of fields of very common use in mathematics. The set
$$\mathbb{Q}_{\sqrt{2}} = \left\{ p + q\sqrt{2} \mid p, q \in \mathbb{Q} \right\}$$

is a field.

Definition 6.2 A **vector space over a field** $\mathbb{F}$ (or linear space) is a set V together with two operations [6–9],
$$+ : V \times V \to V$$

$$\cdot : \mathbb{F} \times V \to V$$

such that

1. Addition is **commutative**:

$$\alpha + \beta = \beta + \alpha \quad \forall \alpha, \beta \in V$$

2. Addition is **associative**:

$$\alpha + (\beta + \gamma) = (\alpha + \beta) + \gamma \quad \forall \alpha, \beta, \gamma \in V$$

3. There is a unique vector $\mathbf{0} \in V$, called the **zero vector** (**neutral additive**), such that

$$\alpha + \mathbf{0} = \mathbf{0} + \alpha = \alpha \quad \forall \alpha \in V$$

4. For each vector $v \in V$ there is a unique vector $-v \in V$ called **inverse additive**, such that

$$v + (-v) = (-v) + v = \mathbf{0}$$

5. $(x \cdot y) \cdot \alpha = x \cdot (y \cdot \alpha) \quad \forall \alpha \in V, \ \forall x, y \in \mathbb{F}$ (**associativity**)
6. $(x + y) \cdot \alpha = x \cdot \alpha + y \cdot \alpha \quad \forall \alpha \in V, \ \forall x, y \in \mathbb{F}$ (**distributivity of scalar multiplication with respect to field addition**).
7. $x \cdot (\alpha + \beta) = x \cdot \alpha + x \cdot \beta \quad \forall \alpha, \beta \in V, \ \forall x \in \mathbb{F}$ (**distributivity of scalar multiplication with respect to vector addition**).
8. **Identity element of scalar multiplication**: $1 \cdot \alpha = \alpha \quad \forall \alpha \in V$

Remark 6.2 The objects of the set V are called *vectors* and the elements of $\mathbb{F}$ are called *scalars*. The subtraction of two vectors and division by a (non-zero) scalar can be defined as

$$\alpha - \beta = \alpha + (-\beta), \quad \forall \alpha, \beta \in V$$
$$\alpha/x = (1/x) \cdot \alpha \quad \forall \alpha \in V, \ \forall x \neq 0, x \in \mathbb{F}$$

Proposition 6.1 *Let V a vector space over a field $\mathbb{F}$, then:*

(I) *If $\beta + \alpha = \gamma + \alpha$ for some $\alpha, \beta, \gamma \in V$ then $\beta = \gamma$.*
(II) $x \cdot \mathbf{0} = \mathbf{0} \quad \forall x \in \mathbb{F}$.
(III) $0 \cdot \alpha = \mathbf{0} \quad \forall \alpha \in V$
(IV) *If $x \cdot \alpha = \mathbf{0}$ for some $x \in \mathbb{F}$ and $\alpha \in V$ then $x = 0$ or $\alpha = \mathbf{0}$.*
(V) $-1 \cdot \alpha = -\alpha \quad \forall \alpha \in V$

Proof According to axioms of a vector space:

1. $\beta = \beta + \mathbf{0} = \beta + (\alpha + (-\alpha)) = (\beta + \alpha) + (-\alpha) = (\gamma + \alpha) + (-\alpha) = \gamma + (\alpha + (-\alpha)) = \gamma + \mathbf{0} = \gamma$.
2. Note that $x \cdot \mathbf{0} = x \cdot \mathbf{0} + \mathbf{0}$ and $x \cdot \mathbf{0} = x \cdot (\mathbf{0} + \mathbf{0}) = x \cdot \mathbf{0} + x \cdot \mathbf{0}$ then $x \cdot \mathbf{0} + x \cdot \mathbf{0} = x \cdot \mathbf{0} + \mathbf{0}$. By I, $x \cdot \mathbf{0} = \mathbf{0}$.
3. In the same manner, $\mathbf{0} + 0 \cdot \alpha = 0 \cdot \alpha = (0 + 0)\alpha = 0 \cdot \alpha + 0 \cdot \alpha$. By I, $0 \cdot \alpha = \mathbf{0}$.

4. By hypothesis, $x \cdot \alpha = \mathbf{0}$. if x the proof is finished. Let suppose that $x \neq 0$ then exists x^{-1} such that $x^{-1}x = 1$ then $\alpha = 1 \cdot \alpha = (x^{-1}x)\alpha = x^{-1}(x\alpha) = x^{-1} \cdot \mathbf{0} = \mathbf{0}$ from II. Hence $\alpha = \mathbf{0}$.
5. Note that $\alpha + (-1)\alpha = 1 \cdot \alpha + (-1)\alpha = (1 + (-1))\alpha = 0 \cdot \alpha = \mathbf{0} = \alpha + (-\alpha)$. Then from I it follows that $-1 \cdot \alpha = -\alpha$.   □

Definition 6.3 Let V a vector space over a field $\mathbb{F}$. A **subspace** of V is a subset nonempty W of V which is itself a vector space over $\mathbb{F}$ with the operations of vector addition and scalar multiplication on V [10–13].

The subspace W is a vector space with the addition and scalar multiplication of $\mathbb{F}$, inherited from V, that is to say, W is closed under the addition and scalar multiplication originally defined in V. To prove that a set W of V is a subspace of V, it suffices to prove:

1. $W \neq \emptyset$.
2. If $w_1, w_2 \in W$ and $x, y \in \mathbb{F}$ then $xw_1 + yw_2 \in W$.

Theorem 6.1 *A nonempty subset W of V is a subset of V if and only if for each pair of vectors $w_1, w_2 \in W$ and each scalar $x \in \mathbb{F}$, the vector $xw_1 + w_2$ is again in W.*

Proof The proof is left to the reader.   □

Definition 6.4 If $W_1, W_2, \ldots, W_k$ are subspaces of V, then the **sum of subspaces** is given by

$$W_1 + W_2 + \cdots + W_k = \{w_1 + w_2 + \cdots + w_k \mid w_1 \in W_1, w_2 \in W_2, \ldots, w_k \in W_k\}$$

Proposition 6.2 *If W_1, W_2 are subspaces of V, then*

1. *$W_1 \cap W_2$ is a subspace of V.*
2. *$W_1 + W_2$ is a subspace of V.*

Proof 1. By hypothesis, W_1, W_2 are subspaces of V then $\mathbf{0} \in W_1$, $\mathbf{0} \in W_2$ so that $\mathbf{0} \in W_1 \cap W2$. Hence $W_1 \cap W2 \neq \emptyset$. Let $x, y \in \mathbb{F}$, $\alpha, \beta \in W_1 \cap W2$, since W_1, W_2 are subspaces of V then $x\alpha + y\beta \in W_1$ and $x\alpha + y\beta \in W_2$ so that $x\alpha + y\beta \in W_1 \cap W_2$. Therefore $W_1 \cap W_2$ is a subspace of V.
2. Since $\mathbf{0} \in W_1, \mathbf{0} \in W_2$ then $\mathbf{0} = \mathbf{0} + \mathbf{0} \in W_1 + W_2$ so that $W_1 + W_2 \neq \emptyset$.   □

Definition 6.5 Let V an vectorial space and W_1, W_2 subspaces of V. It said to be the sum of subspaces is **direct sum** and we write

$$W_1 \oplus W_2$$

to denote the sum $W_1 + W_2$ if

$$W_1 \cap W_2 = \{0\}$$

Proposition 6.3 *Let W_1 and W_2 be subspaces of V. The sum of W_1 and W_2 is direct sum, if and only if every vector v of $W_1 + W_2$ can be written in one and only one way*

$$v = w_1 + w_2, \quad w_1 \in W_1, w_2 \in W_2$$

Proof Suppose that the sum is direct, this is $W_1 \cap W_2 = \{0\}$, let $v \in W_1 + w_2$, then $v = w_1 + w_2 = w_1' + w_2'$ with $w_1, w_1' \in W_1$ and $w_2, w_2' \in W_2$, consider $w_1 + w_2 = w_1' + w_2'$ and $w_1 - w_1' = w_2 - w_2' \in W_1 \cap W_2 = \{0\}$, therefore $w_1 - w_1' = w_2 - w_2' = 0$ leads to $w_1 = w_1'$ and $w_2 = w_2'$.

Any element v of $W_1 + W_2$ can be written as $v = w_1 + w_2$, knowing that $\{0\} \subseteq W_1 \cap W_2$ and using $w \in W_1 \cap W_2$ then $w = w + 0 = 0 + w$ is an element of $W_1 + W_2$, but unicity indicates that $w = 0$, hence
$W_1 + W_2 \subseteq \{0\}$ makes possible to conclude that $W_1 \cap W_2 = \{0\}$. $\qquad\square$

6.2 Generated Subspace

Definition 6.6 Let V a vector space over a field $\mathbb{F}$ and $S \subset V$. The **subspace of V generated by** S denoted by $L(S)$, is the minimum subspace of V containing S, i.e.,

$$L(S) = \bigcap_{W \text{ subspace of } V, S \subset W} W$$

Proposition 6.4 *Let $S \neq \emptyset$. If $S \subset V$ then*

$$L(S) = \{x_1\alpha_1 + x_2\alpha_2 + \cdots + x_n\alpha_n \mid n \in \mathbb{N}, x_i \in \mathbb{F}, \alpha_i \in S, i \in \{1, 2, \ldots, n\}\}$$

Proof Consider $\quad U := \{\alpha_1 v_1 + \cdots + \alpha_n v_n \mid n \in \mathbb{N}, \alpha_1 \in K, v_1 \in S\} \quad$ for $\quad i \in \{1, \ldots, n\}$ then $v = \alpha_1 v_1 + \cdots + \alpha_n v_n \in U$ and W is a subspace of V such that $S \subseteq W$ with $v_1, \ldots, v_n \in S$ makes $v_1, \ldots, v_n \in W$, then $v = \alpha_1 v_1 + \cdots + \alpha_n v_n \in W$. So

$$v \in \bigcap_{W \text{ subspace of } V, S \subset W} W = L(S)$$

Leads to $U \subsetneq L(S)$.

Let $s \in S$ then $s = 1s \in U$ so that $S \subseteq U$, knowing that $S \neq \emptyset$ and $U \neq \emptyset$, makes $\alpha_1 v_1 + \cdots + \alpha_n v_n, \alpha_1' v_1' + \cdots + \alpha_n' v_n' \in U$ and $\alpha, \beta \in K$, Then $\alpha(\alpha_1 v_1 + \cdots + \alpha_n v_n) + \beta(\alpha_1' v_1' + \cdots + \alpha_n' v_n') = \alpha\alpha_1 v_1 + \cdots + \alpha\alpha_n v_n + \beta\alpha_1' v_1' + \cdots + \beta\alpha_n' v_n' \in U$, Thus U is a subspace of V, even more, $S \subseteq U$ makes

$$L(S) = \bigcap_{W \text{ subspace of } V, S \subset W} W \subseteq U$$

Which allows to conclude that $U = L(S)$. $\qquad\square$

Definition 6.7 Let V a vector space over a field $\mathbb{F}$. A vector $\beta \in V$ is said to be **linear combination** of the vectors $\alpha_1, \alpha_2 \ldots, \alpha_n \in V$, it there exist scalars $x_1, x_2 \ldots, x_n \in \mathbb{F}$ such that:

$$\beta = x_1\alpha_1 + x_2\alpha_2 + \cdots + x_n\alpha_n = \sum_{i=1}^{n} x_i\alpha_i$$

Definition 6.8 Let V a vector space over a field $\mathbb{F}$. It is said that vectors $\alpha_1, \alpha_2 \ldots, \alpha_n$ **generate** V (or equivalently that $\{\alpha_1, \alpha_2 \ldots, \alpha_n\}$ **generates** V) if for any $\alpha \in V$ there exist scalars $x_1, x_2 \ldots, x_n \in \mathbb{F}$ such that:

$$\alpha = x_1\alpha_1 + x_2\alpha_2 + \cdots + x_n\alpha_n = \sum_{i=1}^{n} x_i\alpha_i$$

Note that the vectors $\alpha_1, \alpha_2 \ldots, \alpha_n$ generate V if and only if

$$V = L\left(\{\alpha_1, \alpha_2 \ldots, \alpha_n\}\right)$$

6.3   Linear Dependence and Independence

Definition 6.9 Let V a vector space over a field $\mathbb{F}$. It is said that vectors $\alpha_1, \alpha_2 \ldots, \alpha_n$ are **linearly independent over** $\mathbb{F}$ or that $\{\alpha_1, \alpha_2 \ldots, \alpha_n\}$ is **linearly independent over** $\mathbb{F}$ if for $x_1, x_2 \ldots, x_n \in \mathbb{F}$, the equation:

$$x_1\alpha_1 + x_2\alpha_2 + \cdots + x_n\alpha_n = \sum_{i=1}^{n} x_i\alpha_i = 0$$

can only be satisfied by $x_i = 0$ for $i = 1, 2, \ldots, 2$.

This implies that no vector in the set can be represented as a linear combination of the remaining vectors in the set. In other words, a set of vectors is linearly independent if the only representations of **0** as a linear combination of its vectors is the trivial representation in which all the scalars x_i are zero.

It is said that vectors $\alpha_1, \alpha_2 \ldots, \alpha_n$ are **linearly dependent over** $\mathbb{F}$ or that $\{\alpha_1, \alpha_2 \ldots, \alpha_n\}$ is **linearly dependent over** $\mathbb{F}$ if there exists scalars $x_1, x_2 \ldots, x_n \in \mathbb{F}$ *not all zero*, such that

$$x_1\alpha_1 + x_2\alpha_2 + \cdots + x_n\alpha_n = \sum_{i=1}^{n} x_i\alpha_i = 0$$

Notice that if not all of the scalars are zero, then at least one is nonzero, say x_1, in which case this equation can be written in the form

$$\alpha_1 = \frac{-x_2}{x_1}\alpha_2 + \cdots + \frac{-x_n}{x_1}\alpha_n = -\frac{1}{x_1}\sum_{i=2}^{n} x_i\alpha_i$$

Thus, α_1 is shown to be a linear combination of the remaining vectors [3, 8, 9].

Remark 6.3 Finally, note that

1. If $\alpha \in V$, $\alpha \neq 0$ then $\{\alpha\}$ is linearly independent.
2. $\{0\}$ is linearly dependent.
3. $\emptyset$ is linearly independent.
4. If $0 \in \{\alpha_1, \alpha_2 \ldots, \alpha_n\}$ then $\{\alpha_1, \alpha_2 \ldots, \alpha_n\}$ is linearly dependent.

6.4 Bases and Dimension

Definition 6.10 Let V a vector space over a field $\mathbb{F}$. It is said that set of vectors $\{\alpha_1, \alpha_2 \ldots, \alpha_n\}$ of V is **a basis of** V if the set is linearly independent and generates V.

Example 6.2 Let $V = \{a_0 + a_1x + a_2x^2 + a_3x^3 + a_4x^4 \mid a_i \in \mathbb{R}\}$. It is clear that $B = \{1, x, x^2, x^3, x^4\}$ is a basis and dim $V = 5$. Generally, the list $1, x, x^2, x^3, \ldots, x^n$ is a basis of $a_0 + a_1x + a_2x^2 + a_3x^3 + \cdots + a_nx^n$ and dim $V = n + 1$. The dimension of V is the cardinality of a basis B of V. In Chap. 10, this concept will be denoted transcendence degree of an extension field (Definition 10.12).

Lemma 6.1 *The vectors $\{\alpha_1, \alpha_2 \ldots, \alpha_n\}$ nonzero are linearly dependent if and only if on of them is a linear combination of the vectors before him.*

Proof Suppose that α_i is a linear combination of the other vectors:

$$\alpha_i = a_1\alpha_1 + \cdots + a_{i-1}\alpha_{i-1} + a_{i+1}\alpha_{i+1} + \cdots + a_m\alpha_m$$

Adding $-\alpha_i$ to both sides:

$$a_1\alpha_1 + \cdots + a_{i-1}\alpha_{i-1} - \alpha_i + a_{i+1}\alpha_{i+1} + \cdots + a_m\alpha_m = 0$$

With $i \neq 0$, then the vectors are linearly dependent. Suppose that the vectors are linearly dependent:

$$b_1\alpha_1 + \cdots + b_j\alpha_j + \cdots + b_m\alpha_m = 0$$

Having $b_j \neq 0$ leads to the solution:

$$v_j = b_j^{-1} b_1 \alpha_1 - \cdots - b_j^{-1} b_{j-1} \alpha_{j-1} - b_j^{-1} b_{j+1} \alpha_{j+1} - \cdots - b_j^{-1} b_m \alpha_m$$

So α_j is a linear combination of the other vectors. □

Proposition 6.5 *Suppose* $\{\alpha_1, \alpha_2 \ldots, \alpha_n\}$ *generates V and* $\{\beta_1, \beta_2 \ldots, \beta_m\}$ *is linearly independent, then* $m \leq n$.

Proof Since $\{\alpha_i\}$ spans V then $\{\beta_1, \alpha_1, \ldots, \alpha_n\}$ is linearly dependent and spans V, one of the vectors of $\{\beta_1, \alpha_1, \ldots, \alpha_n\}$ is a linear combination of the preceding vectors, but it cannot be β_1 hence it must be any of the other α_i denoted α_j, thus if α_j is removed from the set it is left: $\{\beta_1, \alpha_1, \ldots, \alpha_{j-1}, \alpha_{j+1}, \ldots, \alpha_n\}$, doing the same operation to a vector β_2 makes $\{\beta_1, \beta_2, \alpha_1, \ldots, \alpha_{j-1}, \alpha_{j+1}, \ldots, \alpha_n\}$ which is linearly dependent and spans V. One of the vectors in this last set is a linear combination of the preceding vectors denoted α_k, and removing it from the previous set yields $\{\beta_1, \beta_2, \alpha_1, \ldots, \alpha_{j-1}, \alpha_{j+1}, \alpha_{k-1}, \alpha_{k+1}, \ldots, \alpha_n\}$, if this is done again with a new vector β_3 and so on, and if $m \leq n$ leads to the spanning set of the required form $\{\beta_1, \ldots, \beta_m, \ldots, \alpha_{i1}, \ldots, \alpha_{in-m}\}$, Consider that $m > n$ is not possible, if it where after n operations it will produce $\{\beta_1, \ldots, \beta_m\}$ which implies that β_{n+1} is a linear combination of $\{\beta_1, \ldots, \beta_m\}$ contradicting the hypothesis that $\{\beta_1, \ldots, \beta_m\}$ is linearly independent. □

Corollary 6.1 *Let* $\{\alpha_1, \alpha_2 \ldots, \alpha_n\}$ *and* $\{\beta_1, \beta_2 \ldots, \beta_m\}$ *two basis of V, then* $m = n$.

Proof Suppose $\{\alpha_1, \ldots, \alpha_n\}$ is a basis of V and $\{\beta_1, \beta_2, \ldots\}$ is another basis of V, since $\{\alpha_i\}$ spans V, the basis $\{\beta_1, \beta_2, \ldots\}$ must contain n or less vector, else it is linearly dependent, if the basis $\{\beta_1, \beta_2, \ldots\}$ contains less than n elements, then $\{\alpha_1, \ldots, \alpha_n\}$ is linearly dependent, thus both basis have the same number of elements, hence $m = n$. □

Corollary 6.2 *Let V a vector space that can be generated by a finite number of vectors. Then*

1. *Given any set of generators* $\{\alpha_1, \alpha_2 \ldots, \alpha_n\}$ *of V, it is possible to obtain a basis.*
2. *All linearly independent set* $\{\beta_1, \beta_2 \ldots, \beta_m\}$ *of V can be completed to a basis.*

Proof 1. Remove from $\{\alpha_1, \ldots, \alpha_n\}$ de 0 vectors and successively from left to right the vectors that are linearly dependent from the previous vectors, after this process only a basis remains.

2. let $\{\alpha_1, \ldots, \alpha_n\}$ be a set of generators fo V, consider the set $\{\beta_1, \ldots, \beta_m, \alpha_1, \ldots, \alpha_n\}$ which generates V, removing the 0 vectors and subsequently, from left to right, remove the linearly dependent vectors, after this a basis for V that contains $\{\beta_1, \ldots, \beta_m\}$ is left. □

Definition 6.11 If the vector space V can be generated by a finite number of vectors, we say V is **finite-dimensional**. The number of elements of V is called **the dimension of** V and it is denoted by $\dim V$ or $\dim_{\mathbb{F}} V$ [2, 6, 7].

Proposition 6.6 *Let $\{\alpha_1, \alpha_2 \ldots, \alpha_n\}$ a basis of V. If $\alpha \in V$ then there exist unique $x_1, x_2 \ldots, x_n \in \mathbb{F}$ such that*

$$\alpha = x_1\alpha_1 + x_2\alpha_2 + \cdots + x_n\alpha_n = \sum_{i=1}^{n} x_i\alpha_i$$

Proof The x_i elements exist since $\{\alpha_1, \ldots, \alpha_n\}$ generate V, suppose that

$$v = x_1\alpha_1 + \cdots + x_n\alpha_n$$
$$= y_1\alpha_1 + \cdots + y_n\alpha_n$$

For some $x_1, \ldots, x_n, y_1, \ldots, y_n \in \mathbb{F}$ there is $(x_1 - y_1)\alpha_1 + \cdots + (x_n - y_n)\alpha_n = 0$, since $\{\alpha_1, \ldots, \alpha_n\}$ is linearly independent, then $x_1 - y_1 = \cdots = x_n - y_n = 0$ so $x_n = y_n$, then the values x_i are unique. $\qquad\square$

Definition 6.12 Let V a vector space over a field $\mathbb{F}$, $\{\alpha_1, \alpha_2 \ldots, \alpha_n\}$ a basis of V and $\alpha \in V$. According to Proposition 6.6, there exist unique $x_1, x_2 \ldots, x_n \in \mathbb{F}$ such that $\alpha = \sum_{i=1}^{n} x_i\alpha_i$. The **coordinate vector** of α relative to the base $\{\alpha_1, \alpha_2 \ldots, \alpha_n\}$ is

$$[\alpha]_{\alpha_i} = \begin{pmatrix} x_1 \\ \vdots \\ x_n \end{pmatrix}$$

Proposition 6.7 *Let V a finite-dimensional vector space and W a subspace of V. Then W is finite-dimensional and*

$$dim\, W \leq dim\, V$$

In addition, dim V = dim W if and only if V = W.

Proof 1. Suppose that $dim V = n$, if $W = \{0\}$, then W has finite dimension and $W = 0 \leq n = dimV$.

(i) Suppose that $W \neq \{0\}$, let $w_1 \in W$, $\_w_1 \neq 0$, $\{w_1\}$ is linearly independent in V, then $1 \leq n$. If $L(\{w_1\}) = W$ so $dim W = 1 \leq n = dimV$.

(ii) If $L(\{w_1\}) \subsetneq W$, let $w_2 \in W$, $w_2 \notin L(\{w_1\})$, then $\{w_1, w_2\}$ is linearly independent in V, so $2 \leq n$. Hence $L(\{w_1, w_2\}) = W$, then $dim W = 2 \leq n = dimV$.

(iii) If $L(\{w_1, w_2\}) \subsetneq W$, let $w_3 \in W$, $w_2 \notin L(\{w_1, w_2\})$, then $\{w_1, w_2, w_3\}$ is linearly independent in V, so $3 \leq n$. Hence $L(\{w_1, w_2, w_3\}) = W$, then $dim W = 3 \leq n = dimV$.

This process cannot continue indefinitely, it must stop at n steps, for the r step it is:

(r) Suppose $L(\{w_1, \ldots, w_{r-1}\}) \subsetneq W$, let $w_r \in W$, $w_r \notin L(\{w_1, \ldots, w_{r-1}\})$, then $\{w_1, \ldots, w_r\}$ is linearly independent in V, so $r \leq n$. So $L(\{w_1, \ldots, w_r\}) = W$, then $dim W = r \leq n = dimV$.

2. Clearly $V = W \implies dimV = dimW$, Suppose that $W \subseteq V$ and $dimW = dimV = n$. Let $\{w_1, \ldots, w_n\}$ be a base for W, since $\{w_1, \ldots, w_n\}$ is a subset linearly independent of V and $dimV = n$, then $V = L(\{w_1, \ldots, w_n\}) = W$. $\square$

Theorem 6.2 *Let V a finite-dimensional vector space. Let W_1, W_2 subspaces of V. Then*

$$dim(W_1 + W_2) = dim\,W_1 + dim\,W_2 - dim(W_1 \cap W_2)$$

Proof Let $r = dimW_1 \cap W_2$, $n = dimW_1$ and $m = dimW_2$, let $\{w_1, \ldots, w_r\}$ be a base for $W_1 \cap W_2$, then $\{w_1, \ldots, w_r\}$ is a linearly independent subset of W_1. Complete $\{w_1, \ldots, w_r\}$ making it $\{w_1, \ldots, w_r, u_{r+1}, \ldots, u_n\}$ to be a basis for W_1, it is also completed to $\{w_1, \ldots, w_r, v_{r+1}, \ldots, v_m\}$ so it is a basis for W_2.

1. Suppose that $\alpha_1 w_1 + \cdots + \alpha_r w_r + \beta_{r+1} u_{r+1} + \cdots + \beta_n u_n + \gamma_{r+1} v_{r+1} + \cdots + \gamma_m v_m = 0$ for $\alpha_1 + \cdots + \alpha_r + \beta_{r+1} + \cdots + \beta_n + \gamma_{r+1} + \cdots + \gamma_m \in \mathbb{F}$ it is

$$\alpha_1 w_1 + \cdots + \alpha_r w_r + \beta_{r+1} u_{r+1} + \cdots + \beta_n u_n = -\gamma_{r+1} v_{r+1} - \cdots - \gamma_m v_m \in W_1 \cap W_2$$

Then $-\gamma_{r+1} v_{r+1} - \cdots - \gamma_m v_m = \delta_1 w_1 + \cdots + \delta_r w_r$ for some $\delta_1, \ldots, \delta_r \in \mathbb{F}$. Being $\{w_1, \ldots, w_r, v_{r+1}, \ldots, v_m\}$ a basis for W_2 it is linearly independent, then $\gamma_{r+1} = \cdots = \gamma_m = \delta_1 = cldots = \delta_r = 0$ and $\alpha_1 w_1 + \cdots + \alpha_r w_r + \beta_{r+1} u_{r+1} + \cdots + \beta_n u_n = 0$. Since $\{w_1, \ldots, w_r, u_{r+1}, \ldots, u_n\}$ is a basis for W_1 it is linearly independent, then $\alpha_1 = \cdots = \alpha_r = \beta_{r+1} = \cdots = \beta_n = 0$ makes possible to say that $\{w_1, \ldots, w_r, v_{r+1}, \ldots, v_m\}$ is linearly independent.

2. Let $u + v \in W_1 + W_2$, with $u \in W_1$ and $v \in W_2$, then there is $\alpha_1, \ldots, \alpha_r, \alpha_1', \ldots, \alpha_r', \beta_r, \ldots, \beta_n, \gamma_{r+1}, \ldots, \gamma_m \in \mathbb{F}$ such that

$$u = \alpha_1 w_1 + \cdots + \alpha_r w_r + \beta_{r+1} u_{r+1} + \cdots + \beta_n u_n$$
$$v = \alpha_1' w_1 + \cdots + \alpha_r' w_r + \gamma_{r+1} v_{r+1} + \cdots + \gamma_m v_m$$

Leading to

$$u + v = (\alpha_1 + \alpha_1') w_1 + \cdots + (\alpha_r + \alpha_r') w_r + \beta_{r+1} u_{r+1} + \cdots + \beta_n u_n$$
$$+ \gamma_{r+1} v_{r+1} + \cdots + \gamma_m v_m$$

Therefore $\{w_1, \ldots, w_r, u_{r+1}, \ldots, u_n v_{r+1}, \ldots, v_m\}$ generates $W_1 + W_2$. Points 1 and 2 make possible to conclude that the previous set is a basis for $W_1 + W_2$, Hence

$$dim(W_1 + W_2) = n + m - r$$
$$= dimW_1 + dimW_2 - dim(W_1 \cap W_2) \qquad \square$$

Exercise 6.1 In $\mathbb{R}^3$ proof that the sum of two subspaces of dimension 2 has dimension different than zero.

Corollary 6.3 *If V is a finite-dimensional vector space and $W_1 \cap W_2 = \{0\}$ then*

$$dim(W_1 \oplus W_2) = dim\,W_1 + dim\,W_2$$

Proof Since $W_1 \cap W_2 = \{0\}$ the direct sum and the dimension yields $dim\,(W_1 \cap W_2)$ = 0, from the theorem it follows that $dim\,(W_1 \oplus W_2) = dim\,W_1 + dim\,W_2$. □

6.5 Quotient Space

Let V a vector space over a field $\mathbb{F}$ and W a subspace of V. Let $\alpha, \beta \in V$. It said to be α and β are **related** or **related module** W and we write $\alpha \sim \beta$, if $\alpha - \beta \in W$. This relation is an equivalence relation. Indeed:

1. Let $\alpha \in V$, then $\alpha - \alpha = \mathbf{0} \in W$, therefore $\alpha \sim \alpha$.
2. Supposing $\alpha \sim \beta$, then $\alpha - \beta \in W$. Finally $\beta - \alpha = -(\alpha - \beta) \in W$. Therefore $\beta \sim \alpha$.
3. Supposing $\alpha \sim \beta$ and $\beta \sim \gamma$, then $\alpha - \beta \in W$ and $\beta - \gamma \in W$. Now, $\alpha - \gamma = \alpha - \beta + \beta - \gamma \in W$. Hence $\alpha \sim \gamma$.

According to Definition 1.16, the quotient set is denoted by V/W. The elements of V/W are the equivalence classes of the elements of V. Let $\alpha \in V$, then:

$$
\begin{aligned}
[\alpha] &= \{\beta \in V \mid \beta \sim \alpha\} \\
&= \{\beta \in V \mid \beta - \alpha \in W\} \\
&= \{\beta \in V \mid \beta - \alpha = \gamma \in W\} \\
&= \{\alpha + \gamma \mid \gamma \in W\} \\
&= \alpha + W
\end{aligned}
$$

These classes form a partition of V. It defined in V/W an addition and scalar multiplication. These operations are well defined:

1. $(\alpha + W) + (\beta + W) = (\alpha + \beta) + W, \quad \forall\, \alpha, \beta \in V$
2. $x\,(\alpha + W) = (x\alpha) + W, \quad \forall\, \alpha \in V, \forall\, x \in \mathbb{F}$

To verify that operations are well defined, note the following operations. Supposing that $\alpha + W = \alpha_1 + W$ and $\beta + W = \beta_1 + W$, then $\alpha - \alpha_1 \in W$, $\beta - \beta_1 \in W$ and $(\alpha + \beta) - (\alpha_1 + \beta_1) = (\alpha - \alpha_1) + (\beta - \beta_1) \in W$. Therefore $(\alpha + \beta) + W = (\alpha_1 + \beta_1) + W$. Similarly for the multiplication, supposing $\alpha + W = \alpha_1 + W$ then $\alpha - \alpha_1 \in W$. Let $x \in \mathbb{F}$, then $x\alpha - x\alpha_1 = x(\alpha - \alpha_1) \in W$. Therefore $x\alpha + W = x\alpha_1 + W$. We can conclude that V/W is a vector space over $\mathbb{F}$ with operations defined above. This vector space is called the **quotient space**.

Remark 6.4 The form of the quotient ring associated with its ideal is important to define the differential range of output given in Chap. 10.

The elements of quotient space have the form $\alpha + W, \alpha \in W$. The neutral for addition is the class of 0, i.e., $0 + W$. The inverse for addition (the inverse of $\alpha + W$) is $(-\alpha) + W$.

Lemma 6.2 *If V is a finite-dimensional vector space and W is a subspace of V then* *dim V/W is finite. In addition:*

$$dim\ V/W \leq dim\ V$$

Proof Having the finite set $\{v_1, \ldots, v_n\}$ that generates V, the set $\{v_1 + W, \ldots, v_n + W\}$ generates V/W which has finite dimension, if $\{v_1, \ldots, v_n\}$ is a basis for V, then $\{v_1 + W, \ldots, v_n + W\}$ generates V/W, then $V/W \leq dim$V. □

Proposition 6.8 *Let V a finite-dimensional vector space over a field $\mathbb{F}$ and W a* *subspace of V. Then*
$$dim\ V = dim\ W + dim\ V/W$$

Proof Let $n = dim V$, W has a finite dimension $r = dim W \leq n$, the lemma states V/W has finite dimension $s = dim V/W \leq n$, having $\{w_1, \ldots, w_r\}$ be a base for W and $\{u_1 + W, \ldots, u_s + W\}$ is a base for V/W, then $\{w_1, \ldots, w_r, u_1, \ldots, u_n\}$ is a base for V.

(1) Consider $v \in V$, since $v + W \in V/W$ and $\{u_1 + W, \ldots, u_s + W\}$ is a basis for V/W, there is $\alpha_1, \ldots, \alpha_s \in \mathbb{F}$ such that $v + W = \alpha_1 (u_1 + W) + \cdots + \alpha_s (u_s + W)$, then $v + W = (\alpha_1 u_1 + \cdots + \alpha_s u_s) + W$ so $v - (\alpha_1 u_1 + \cdots + \alpha_s u_s) = w \in W$.

Since $\{w_1, \ldots, w_r\}$ is a basis for W, there are $\beta_1, \ldots, \beta_r \in \mathbb{F}$ such that $w = \beta_1 w_1 + \cdots + \beta_r w_r$, therefore

$$v = w + \alpha_1 u_1 + \cdots + \alpha_s u_s$$
$$= \beta_1 w_1 + \cdots + \beta_r w_r + \alpha_1 u_1 + \cdots + \alpha_s u_s$$

The set $\{w_1, \ldots, w_r, u_1, \ldots u_n\}$ generates V.

(2) Suppose that $\delta_1 w_1 + \cdots + \delta_r w_r + \gamma_1 u_1 + \cdots + \gamma_s u_s = 0$ for some $\delta_1, \ldots, \delta_r, \gamma_1, \ldots, \gamma_s \in \mathbb{F}$, then

$$\delta_1 (w_1 + W) + \cdots + \delta_r (w_r + W) + \gamma_1 (u_1 + W) + \cdots + \gamma_1 (u_1 + W) = 0 + W$$

Makes $\gamma_1 (u_1 + W) + \cdots + \gamma_1 (u_1 + W) = 0 + W$, since $\{u_1 + W, \ldots, u_s + W\}$ is a basis for V/W it is linearly independent in $\mathbb{F}$, so $\gamma_1 = \ldots = \gamma_s = 0$, then $\delta_1 w_1 + \cdots + \delta_r w_r = 0$, since $\{w_1, \ldots, w_r\}$ is a basis for W and is linearly independent causes $\delta_1 = \ldots = \delta_r = 0$ so $\{w_1, \ldots, w_r, u_1, \ldots, u_s\}$ is also linearly independent. From 1 and 2 it is possible to say that $\{w_1, \ldots, w_r, u_1, \ldots, u_s\}$ is a basis for V, then $dim V = r + s = dim W + dim V/W$. □

6.6 Cayley-Hamilton Theorem

Theorem 6.3 *Every matrix is a zero of its characteristic polynomial.*
To illustrate this theorem, we have the following example:

$$A = \begin{bmatrix} 1 & 2 \\ 3 & 2 \end{bmatrix}$$

The characteristic polynomial of the matrix is:

$$\begin{aligned} |\lambda I - A| &= |A - \lambda I| \\ &= \begin{bmatrix} 1-\lambda & 2 \\ 3 & 2-\lambda \end{bmatrix} \\ &= (1-\lambda)(2-\lambda) - 6 \\ &= \lambda^2 - 3\lambda - 4 \\ &= P(\lambda) \end{aligned}$$

A is a zero for $P(\lambda)$

$$P(A) = \begin{bmatrix} 1 & 2 \\ 3 & 2 \end{bmatrix}^2 - 3\begin{bmatrix} 1 & 2 \\ 3 & 2 \end{bmatrix} - 4\begin{bmatrix} 1 & 0 \\ 0 & 1 \end{bmatrix}$$
$$P(A) = 0$$

In what follows we give the proof of Theorem 6.3.

Proof Let A be a squared matrix of size $n \times n$ and $P(\lambda)$ is a characteristic polynomial:

$$P(A) = |A - \lambda I| = \lambda^n + a_{n-1}\lambda^{n-1} + \cdots + a_1\lambda + a_0$$

Suppose that $B(\lambda)$ represents the adjoint matrix of $\lambda I - A$. The elements of $B(\lambda)$ are the cofactors of the matrix $\lambda I - A$ hence they are polynomials in λ of order no larger than $n - 1$:

$$B(\lambda) = B_{n.1}\lambda^{n-1} + B_{n-2}\lambda^{n-2} + \cdots + B_1\lambda + B_0$$

Where $B_i, 0 \le i \le n - 1$ are the squared matrices of size $n \times n$ over the field K. Using the fundamental property of the adjoint:

$$(\lambda I - A) B(\lambda) = |\lambda I - A| I$$
$$\frac{B(\lambda)}{|\lambda I - A|} = (\lambda I - A)^{-1}$$

This yields to:

$$(\lambda I - A)\left(B_{n-1}\lambda^{n-1} + \cdots + B_1\lambda + B_0\right) = \left(\lambda^n + a_{n-1}\lambda^{n-1} + \cdots + a_1\lambda + a_0\right) I$$

Then:

$$B_{n-1} = I\left(\lambda^n\right)$$
$$B_{n-2} - AB_{n-1} = a_{n-1}I\left(\lambda^{n-1}\right)$$
$$B_{n-3} - AB_{n-2} = a_{n-2}I\left(\lambda^{n-2}\right)$$

$$\vdots \quad \vdots \quad \vdots$$

$$B_0 - AB_1 = a_1I\left(\lambda\right)$$
$$-AB_0 = a_oI$$

Multiplying by $A^n, A^{n-1}, \ldots . A, I$:

$$A^n B_{n-1} = A^n$$
$$A^{n-1}B_{n-2} - A^n B_{n-1} = a_{n-1}A^{n-1}$$
$$A^{n-2}B_{n-3} - A^{n-1}B_{n-2} = a_{n-n}A^{n-2}$$

$$\vdots \quad \vdots \quad \vdots$$

$$AB_0 - A^2B_1 = a_1A$$
$$-AB_0 = a_oI$$

Adding the matrix equations to the right side:

$$P\left(A\right) = A^n + a_{n-1}A^{n-1} + a_{n-2}A^{n-2} + \cdots + a_1A + a_0I = 0$$

Hence, the Cayley-Hamilton theorem is true. $\square$

6.7   Euclidean Spaces

Definition 6.13 Let V a vector space over $\mathbb{R}$. An **inner product** in V is a function $\langle \cdot, \cdot \rangle : V \times V \longrightarrow \mathbb{R}$ such that:

1. $\langle x_1\alpha_1 + x_2\alpha_2, \beta \rangle = x_1\langle \alpha_1, \beta \rangle + x_2\langle \alpha_2, \beta \rangle, \forall\, \alpha_1, \alpha_2, \beta \in V, \forall\, x_1, x_2 \in \mathbb{R}$.
2. $\langle \alpha, \beta \rangle = \langle \beta, \alpha \rangle, \forall\, \alpha, \beta \in V$.
3. $\langle \alpha, \alpha \rangle \geq 0, \forall\, \alpha \in V$ and $\langle \alpha, \alpha \rangle = 0$ if and only if $\alpha = 0$.

Definition 6.14 Let V a vector space over $\mathbb{C}$. An **inner product** in V is a function $\langle \cdot, \cdot \rangle : V \times V \longrightarrow \mathbb{C}$ such that:

1. $\langle x_1 \alpha_1 + x_2 \alpha_2, \beta \rangle = x_1 \langle \alpha_1, \beta \rangle + x_2 \langle \alpha_2, \beta \rangle, \forall \, \alpha_1, \alpha_2, \beta \in V, \forall \, x_1, x_2 \in \mathbb{C}$.
2. $\langle \alpha, \beta \rangle = \overline{\langle \beta, \alpha \rangle}, \forall \, \alpha, \beta \in V$.
3. $\langle \alpha, \alpha \rangle \geq 0, \forall \, \alpha \in V$ and $\langle \alpha, \alpha \rangle = 0$ if and only if $\alpha = 0$.

Remark 6.5 An **inner product space** is a vector space V over a field $\mathbb{F}$ with inner product. The inner product space is also known as a real inner product space or Euclidean space if $\mathbb{F} = \mathbb{R}$. If $\mathbb{F} = \mathbb{C}$ the inner product space is called complex inner product space or unitarian space.

Definition 6.15 Let V an Euclidean space (or unitarian space) and $\alpha \in V$. The **norm** of u is given by:
$$\|\alpha\| = \sqrt{\langle \alpha, \alpha \rangle}$$

Definition 6.16 A vector $\alpha \in V$ is said to be **unitarian** if $\|\alpha\| = 1$.

Let $\alpha \in V, \alpha \neq 0$. To **normalize** α, we consider the unitarian vector
$$\alpha_N = \frac{\alpha}{\|\alpha\|} = \frac{1}{\|\alpha\|} \cdot \alpha$$

Indeed:
$$\begin{aligned}
\|\alpha_N\|^2 &= \langle \alpha_N, \alpha_N \rangle \\
&= \langle \frac{1}{\|\alpha\|} \cdot \alpha, \frac{1}{\|\alpha\|} \cdot \alpha \rangle \\
&= \frac{1}{\|\alpha\|} \langle \alpha, \frac{1}{\|\alpha\|} \cdot \alpha \rangle \\
&= \frac{1}{\|\alpha\|^2} \langle \alpha, \alpha \rangle \\
&= \frac{1}{\|\alpha\|^2} \|\alpha\|^2 \\
\|\alpha_N\|^2 &= 1
\end{aligned}$$

Theorem 6.4 (Cauchy-Schwarz inequality) *Let V an Euclidian (or unitarian) space.*
$$|\langle \alpha, \beta \rangle| \leq \|\alpha\| \cdot \|\beta\| \, \forall \alpha, \beta \in V \qquad \square$$

Let V an Euclidian space. According to Cauchy-Schwarz inequality, in $\mathbb{R}^2$, for $\alpha, \beta \neq 0$:
$$\cos \theta = \frac{\langle \alpha, \beta \rangle}{\|\alpha\| \cdot \|\beta\|} \qquad (6.1)$$

Definition 6.17 Let V an Euclidian space and $\alpha, \beta \in V$. It is said that vectors α and β are **orthogonal** if

$$\langle \alpha, \beta \rangle = 0$$

Note that

$$\langle 0, \beta \rangle = \langle 0 \cdot 0, \beta \rangle = 0 \langle 0, \beta \rangle = 0 \; \forall \beta \in V$$

Definition 6.18 Let V an Euclidian space and W a subspace of V. The **orthogonal complement** of W is:

$$W^\mathsf{T} = \{ \alpha \in V \mid \langle \alpha, \beta \rangle = 0 \; \forall \beta \in W \}$$

Proposition 6.9 *Let V an Euclidian space and W a subspace of V. W^T is a subspace of V.*

Proof It is left to the reader as an exercise. □

Definition 6.19 Let V an Euclidian space [1, 6, 13].

1. A set of vectors $\{\alpha_1, \ldots, \alpha_r\}$ of V is said to be **orthogonal** if its elements are orthogonal pairs, i.e.:

$$\langle \alpha_i, \alpha_j \rangle = 0 \quad \text{if } i \neq j$$

2. A set of vectors $\{\alpha_1, \ldots, \alpha_r\}$ of V is said to be **orthonormal** if its elements are orthonormal pairs, i.e.:

$$\langle \alpha_i, \alpha_j \rangle = \delta_{ij}$$

6.8 GramSchmidt Process

The GramSchmidt process is a method for the orthonormalization of a set of vectors in an inner product space, most commonly the Euclidean space $\mathbb{R}^n$.

Lemma 6.3 *Let V an Euclidian space. Let $\{\alpha_1, \ldots, \alpha_n\}$ an orthogonal set of vectors of V, with $\alpha_i \neq 0 \; \forall 1 \leq i \leq n$. Then*

1. *$\{\alpha_1, \ldots, \alpha_n\}$ is linearly independent.*
2. *If $\beta \in V$, then γ, the vector given by*

$$\gamma = \beta - \frac{\langle \beta, \alpha_1 \rangle}{\| \alpha_1 \|^2} \alpha_1 - \frac{\langle \beta, \alpha_2 \rangle}{\| \alpha_2 \|^2} \alpha_2 - \ldots - \frac{\langle \beta, \alpha_n \rangle}{\| \alpha_n \|^2} \alpha_n$$

is an orthogonal vector to $\alpha_i, \; \forall 1 \leq i \leq n$.

Proof It is left to the reader as an exercise. □

The GramSchmidt process described below, is an algorithm such that, given a basis $\{\alpha_1, \ldots, \alpha_n\}$ of an euclidian space V and for the Lemma 6.3, it produces an orthogonal basis $\{\beta_1, \ldots, \beta_n\}$ of V. In addition, the change of basis matrix is superior triangular:

Theorem 6.5 *Let V a finite dimensional euclidian space and W a subspace of V, then*

$$V = W \oplus W^\top$$

Proof It is left to the reader as an exercise. $\qquad\square$

References

1. Hoffman, K., Kunze, R.: Linear Algebra. Prentice-Hall (1971)
2. Grossman, S.: Elementary Linear Algebra, Wadsworth Publishing (1987)
3. Nef, W.: Linear Algebra. Dover Publications (1967)
4. Nering, E.: Linear Algebra and Matrix Theory. Wiley (1970)
5. Gantmacher, F.R.: The Theory of Matrices, vol. 1. Chelsea, New York (1959)
6. Axler, S.: Linear Algebra Done-Right. Springer (1997)
7. Lipschutz, S.: Linear Algebra. McGraw Hill (2009)
8. Whitelaw, T.: Introduction to Linear Algebra, Chapman & Hall/CRC (1992)
9. Gantmacher, F.R.: The Theory of Matrices, vol. 2. Chelsea, New York (1964)
10. Lang, S.: Linear Algebra. Springer (2004)
11. Anton, H.: Elementary Linear Algebra. Wiley (2010)
12. kolman, B.: Introductory Linear Algebra an Applied First Course. Pearson Education (2008)
13. Halmos, P.R.: Finite-dimensional Vector Spaces. Courier Dover Publications (2017)

Chapter 7
Linear Transformations

Abstract This chapter is about linear transformations, it begins with an elementary introduction, followed by kernel and image concepts, then linear operators and finally associate matrix.

7.1 Background

Definition 7.1 Let V and W vector spaces over a field $\mathbb{F}$. A **linear transformation** [1–5] between two vector spaces V and W is a map:

$$T : V \longrightarrow W$$

such that the following hold:

1. $T(\alpha_1 + \alpha_2) = T(\alpha_1) + T(\alpha_2) \ \forall \alpha_1, \alpha_2 \in V$
2. $T(x\,\alpha_1) = x\,T(\alpha_1) \ \forall x \in \mathbb{F}$.

Remark 7.1 The above conditions, can be substituted by the following condition:

$$T(x_1\alpha_1 + x_2\alpha_2) = x_1 T(\alpha_1) + x_2\,T(\alpha_2) \ \forall \alpha_1, \alpha_2 \in V, \forall x_1, x_2 \in \mathbb{F}$$

Proposition 7.1 *Let V, W, U vector spaces over a field $\mathbb{F}$. Let $T_1 : V \longrightarrow W$, $T_2 : W \longrightarrow U$ linear transformations. The composition $T_2 \circ T_1 : V \longrightarrow U$ is a linear transformation.*

Proof Let $\alpha_1, \alpha_2 \in V$ and $x_1, x_2 \in \mathbb{F}$. Then, we have:

© Springer Nature Switzerland AG 2019
R. Martínez-Guerra et al., *Algebraic and Differential Methods for Nonlinear Control Theory*, Mathematical and Analytical Techniques with Applications to Engineering, https://doi.org/10.1007/978-3-030-12025-2_7

$$(T_2 \circ T_1)(x_1\alpha_1 + x_2\alpha_2) = T_2(T_1(x_1\alpha_1 + x_2\alpha_2))$$
$$= T_2(x_1 T_1(\alpha_1) + x_2 T_1(\alpha_2))$$
$$= x_1 T_2(T_1(\alpha_1)) + x_2 T_2(T_1(\alpha_2))$$
$$= x_1(T_2 \circ T_1)(\alpha_1) + x_2(T_2 \circ T_1)(\alpha_2)$$

Therefore $T_2 \circ T_1 : V \longrightarrow U$ is a linear transformation. $\qquad\square$

Remark 7.2 A linear transformation may or may not be injective or surjective [6–8]. If V and W have the same dimension, then it is possible for T to be invertible, i.e., there exists a T^{-1} such that $T T^{-1} = I$.

7.2   Kernel and Image

Proposition 7.2 *Let $T : V \longrightarrow W$ a linear transformation, then $T(0) = 0$.*

Proof Since $0 + T(0) = T(0) = T(0 + 0) = T(0) + T(0)$, then $T(0) = 0$. $\qquad\square$

Definition 7.2 Let $T : V \longrightarrow W$ a linear transformation. The **kernel** of T is given by the following set [9–12]:

$$\ker T = \{\alpha \in V \mid T(\alpha) = 0\}$$

Proposition 7.3 *Let $T : V \longrightarrow W$ a linear transformation. The kernel of T is a subspace of V.*

Proof 1. Knowing that $T(0) = 0$, then $0 \in \ker T$ and $\ker T \neq 0$
   2. Let $v_1, v_2 \in V$ and $\alpha_1, \alpha_2 \in \mathbb{F}$, make $T(\alpha_1 v_1 + \alpha_2 v_2) = \alpha_1 T(v_1) + \alpha_2 T(v_2) = \alpha_1 \cdot 0 + \alpha_2 \cdot 0 = 0$,
   then $\alpha_1 v_1 + \alpha_2 v_2 \in \ker T$ suggests that the kernel of T is a subspace of V. $\qquad\square$

Proposition 7.4 *Let $T : V \longrightarrow W$ a linear transformation. T is injective if and only if $\ker T = \{0\}$.*

Proof 1. Suppose T is injective, then $\{0\} \subseteq \ker T$ and $v \in \ker T$, so $T(v) = 0 = T(0)$, since T is injective, $v = 0$, where $\ker T \subseteq \{0\}$, hence $\ker T = \{0\}$.
   2. Suppose $\ker T = \{0\}$ and $v_1, v_2 \in V$ such that $T(v_1) = T(v_2)$ then $T(v_1) = T(v_2) \implies T(v_1) - T(v_2) = 0 \implies T(v_1 + v_2) = 0 \implies v_1 - v_2 \in 1 \in \ker T \implies v_1 - v_2 = 0 \implies v_1 = v_2$ then T is injective. $\qquad\square$

Proposition 7.5 *Let $T : V \longrightarrow W$ a linear transformation. The **image** of T defined by:*

$$ImT = \{T(\alpha) \mid \alpha \in V\}$$

is a subspace of W.

Proof 1. Knowing that $0 = T(0) \in ImT$ and $0 \in ImT$ so $ImT \neq \emptyset$

2. Let $T(\alpha_1), T(\alpha_2) \in ImT$ and $v_1, v_2 \in \mathbb{F}$, having $v_1 T(\alpha_1) + v_2 T(\alpha_2) = T(\alpha_1 v_1 + \alpha_2 v_2) \in ImT$ makes possible to conclude that the image of T is a subspace of W. $\square$

Proposition 7.6 *Let V a vector space over a field $\mathbb{F}$ and W a subspace of V. The function $\pi : V \longrightarrow V/W$, defined by $\pi(\alpha) = \alpha + W$, with $\alpha \in V$ is a linear transformation. In addition, π is surjective.*

Proof 1. Let $\alpha_1, \alpha_2 \in V$ and $v_1, v_2 \in \mathbb{F}$, having $\pi(v_1\alpha_1 + v_2\alpha_2) = (v_1\alpha_1 + v_2\alpha_2) + W = (v_1\alpha_1 + W) + (v_2\alpha_2 + W) = v_1(\alpha_1 + W) + v_2(\alpha_2 + W) = v_1\pi(\alpha_1) + v_2\pi(\alpha_2)$, then π is a linear transformation.

2. Given that $\alpha + W \in V/W$, then $\alpha \in V$ equals $\pi(\alpha) = \alpha + W$, then π is surjective. $\square$

Theorem 7.1 *Let V a finite-dimensional vector space over a field $\mathbb{F}$ and $T : V \longrightarrow U$ a linear transformation. Then*

$$dimV = dim\,kerT + dim\,ImT$$

Proof Being $n = \dim V$, since $\ker T$ is a subspace of V, $\ker T$ has a finite dimension $r := \dim \ker T \leq n$. Let $\{w_1, \ldots, w_r\}$ be a base for $\ker T$ and a subset of V linearly independent over $\mathbb{F}$. Completing $\{w_1, \ldots, w_r\}$ to be a base of V: $\{w_1, \ldots, w_r, v_1, \ldots, v_{n-r}\}$:

1. Let $T(v) \in ImT$, with $v \in V$, there is $\alpha_1, \ldots, \alpha_r, \beta_1, \ldots, \beta_{n-r} \in \mathbb{F}$ such that $v = \alpha_1 w_1 + \cdots + \alpha_r w_r + \beta_1 v_1 + \cdots + \beta_{n-r} v_{n-r}$, then $T(v) = \beta_1 T(v_1) + \cdots + \beta_{n-r} T(v_{n-r})$ and $\{T(v_1), \ldots, T(v_{n-r})\}$ generates ImT .

2. Consider $\{T(v_1), \ldots, T(v_{n-r})\}$ then, $\gamma_1 T(v_1) + \cdots + \gamma_{n-r} T(v_{n-r}) = 0$ for some $\gamma_1, \ldots, \gamma_{n-r} \in \mathbb{F}$. Having T being linearly independent $T(\gamma_1 v_1 + \cdots + \gamma_{n-r} v_{n-r}) = 0$, then $\gamma_1 v_1 + \cdots + \gamma_{n-r} v_{n-r} \in \ker T$, in consequence there exist $\delta_1, \ldots, \delta_r \in \mathbb{F}$ such that $\gamma_1 v_1 + \cdots + \gamma_{n-r} v_{n-r} = \delta_1 w_1 + \cdots + \delta_r w_r$, so $-\delta_1 w_1 - \cdots - \delta_r w_r + \gamma_1 v_1 + \cdots + \gamma_{n-r} v_{n-r} = 0$, knowing that $\{w_1, \ldots, w_r, v_1, \ldots, v_{n-r}\}$ is a basis for V, it is linearly independent, making $-\delta_1 = \cdots = -\delta_r = \gamma_1 = \cdots = \gamma_{n-r} = 0$ so $\{T(v_1), \ldots, T(v_{n-r})\}$ is linearly independent.

Therefore $\{T(v_1), \ldots, T(v_{n-r})\}$ is a basis for ImT and $\dim V = n = r + n - r = \dim \ker T + \dim ImT$. $\square$

In what follows we give a direct application of Theorem 7.1 to linear control theory, in particular of the concept of observability.

The next figure shows a object with mass m that is displaced without friction, to which a force F is applied producing an acceleration a (Fig. 7.1).

Fig. 7.1 System diagram

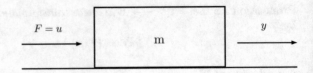

Example 7.1 The mentioned system diagram is as follows:
From Newton's second law we have:

$$F = ma$$
$$= m\frac{d^2y}{dt}$$

Here, $\frac{dy}{dt} = y' = \dot{y} = y^{(1)}$, $\frac{d^2y}{dt} = y'' = \ddot{y} = y^{(2)}$ are commonly used to represent the first and second derivatives with respect to the time respectively.

Now, considering $m = 1$

$$F = \frac{d^2y}{dt}$$

And the input $F = \frac{d^2y}{dt} = u$.

Now, we consider the following change of coordinates:

$$x_1 = y$$
$$x_2 = \dot{y}$$

Giving the following state representation:

$$\dot{x}_1 = x_2$$
$$\dot{x}_2 = u$$

With

$$x = \begin{bmatrix} x_1 \\ x_2 \end{bmatrix} \in \mathbb{R}^2$$

And

$$\dot{x} = \begin{bmatrix} \dot{x}_1 \\ \dot{x}_2 \end{bmatrix} \in \mathbb{R}^2$$

The system can be expressed as:

$$\begin{bmatrix} \dot{x}_1 \\ \dot{x}_2 \end{bmatrix} = \begin{bmatrix} 0 & 1 \\ 0 & 0 \end{bmatrix} \begin{bmatrix} x_1 \\ x_2 \end{bmatrix} + \begin{bmatrix} 0 \\ 1 \end{bmatrix} u$$

$$y = \begin{bmatrix} 0 & 1 \end{bmatrix} \begin{bmatrix} x_1 \\ x_2 \end{bmatrix}$$

That is to say

$$\dot{x} = Ax + Bu$$
$$y = Cx$$
$$x(t_0) = x_o$$

For $x \in \mathbb{R}^n$, $u \in \mathbb{R}^m$, the matrices A, B, C are of appropriate size.
It is well known that a linear system is controlable and observable if the rank of $\begin{bmatrix} B & AB \end{bmatrix} = 2$ that is to say the system is fully controllable, as well as the rank $\begin{bmatrix} C & CA \end{bmatrix}^T = 2$ that is to say that the system is fully observable (see [13]).

Definition 7.3 (*Distinguishability*) Two initial states x_0 and x_1 are distinguishable for an input u if the produced outputs by both initial states are different, and if they are not distinguishable they are indistinguishable [14].

Exercise 7.1 Show that the indistinguishability is a equivalence relation (Chap. 1, Definition 1.9).

Definition 7.4 A system is said to be observable if all of its states are distinguishable.

Remark 7.3 It is clear that the injectivity of the output function is linked to the observability concept.

Theorem 7.2 ([15]) *A system is fully observable if and only if the rank of the observability matrix* $V = \begin{bmatrix} C \\ CA \\ \vdots \\ CA^{n-1} \end{bmatrix}$ *is equal to the dimension of the space state.*

Proof (Sufficiency) By contradiction, suppose that the system is not fully observable. If the system is not observable then, there exists states x_0 and x_1 with $x_0 \neq x_1$ $\forall u, \forall t > 0$ such that $y(x_0, t, u) = y(x_1, t, u)$.
    Let

$$\dot{x} = Ax + Bu$$
$$y = Cx$$

Then

$$\dot{x} - Ax = Bu$$
$$\Rightarrow e^{-At} (\dot{x} - Ax) = e^{-At} (Bu)$$
$$\Rightarrow \left(e^{-At} x \right)^{\cdot} = e^{-At} Bu$$
$$\Rightarrow e^{-At} x (\tau) \mid_0^t = \int_0^t e^{-A\delta} Bu (\delta) \, d\delta$$

Hence

$$x = e^{At} \int_0^t e^{-A\delta} Bu (\delta) \, d\delta + x (0) e^{At},$$

Since $y = Cx$, we have

$$y (x_0) = Ce^{At} x_0 + \int_0^t e^{A(t-\delta)} C Bu (\delta) \, d\delta, \quad x (0) = x_0$$
$$y (x_1) = Ce^{At} x_1 + \int_0^t e^{A(t-\delta)} C Bu (\delta) \, d\delta, \quad x_1 (0) = x_1$$

Since $y (x_0) = y (x_1)$
This yields to

$$Ce^{At} (x_1 - x_0) = 0, \quad with \quad x_1 \neq x_0$$

i.e.

$$0 = \begin{cases} Ce^{At} (x_1 - x_0) \\ CAe^{At} (x_1 - x_0) \\ \quad \vdots \\ CA^{n-1} e^{At} (x_1 - x_0) \end{cases}$$

We define $x = x_1 - x_0 \neq 0$
That is to say

$$\begin{bmatrix} Cx \\ CAx \\ \\ CA^{n-1}x \end{bmatrix} e^{At} = 0, \quad for \quad e^{At} \neq 0$$

Then

$$x \in \ker \begin{bmatrix} Cx \\ CAx \\ \\ CA^{n-1}x \end{bmatrix} \implies \dim \ker \begin{bmatrix} Cx \\ CAx \\ \\ CA^{n-1}x \end{bmatrix} \neq 0$$

From Theorem 7.1 $\dim V = \dim Im(V) + \dim \ker(V)$ since $\dim V = n$, then $\dim Im(V) < n$ and, in consequence $dim Im(V) = rank(V) < n$, this ends the sufficiency.

The necessity is left to the reader as an exercise. $\square$

Definition 7.5 Let V and W vector spaces over the field $\mathbb{F}$. An **isomorphism** $T : V \longrightarrow W$ is a bijective linear transformation.

Proposition 7.7 *Let $T : V \longrightarrow W$ an isomorphism, then $T^{-1} : W \longrightarrow V$ is an isomorphism.*

Proof Since that $T^{-1} : W \to V$ is a bijective function and $T^{-1}(w) = v$ if and only if $T(v) = w$, let $w_1, w_2 \in W$, $\alpha_1, \alpha_2 \in \mathbb{F}$, and $v_1, v_2 \in V$ such that $v_1 = T(w_1)$ and $v_2 = T(w_2)$, since T is linear, there is $T(\alpha_1 v_1 + \alpha_2 v_2) = \alpha_1 T(v_1) + \alpha_2 T(v_2)$, then $T^{-1}(\alpha_1 w_1 + \alpha_2 w_2) = \alpha_1 v_1 + \alpha_2 v_2 = \alpha_1 T^{-1}(w_1) + \alpha_2 T^{-1}(w_2)$ allows to conclude that T^{-1} is a linear transformation and it is also an isomorphism. $\square$

Definition 7.6 Let V and W vector spaces over the field $\mathbb{F}$. It is said that V and W are isomorphic if there exists an isomorphism of V in W. We write $V \cong W$, and read V is isomorphic to W, when such a map exists.

Theorem 7.3 *Let V and W finite-dimensional vector spaces over a field $\mathbb{F}$. Then $V \cong W$ if and only if $dim V = dim W$.*

Proof 1. Suppose that V and W are isomorphisms, there exists the isomorphism $T : V \to W$, let $\{v_1, \ldots, v_n\}$ be a base for V, then it is linearly independent and generates V. Being T injective and the sets $\{v_1, \ldots, v_n\}, \{T(v_1), \ldots, T(v_n)\}$ are linearly independent. Since T is surjective and $\{v_1, \ldots, v_n\}$ generates V, $\{T(v_1), \ldots, T(v_n)\}$ generates W, then $\dim W = n = \dim V$.

2. Suppose that $\dim V = \dim W = n$, let $\{v_1, \ldots, v_n\}$ and $\{w_1, \ldots, w_n\}$ be bases for V and W, defining $T : V \to W$ for $T(v_i) = w_i$ for $1 \leq i \leq n$, T is extended by linearity, this is:

(a) Let $v \in \ker T$, sincet there are unique $\alpha_1, \ldots, \alpha_n \in \mathbb{F}$ such that $v = \alpha_1 v_1 + \cdots + \alpha_n v_n$ and $T(v) = 0$, then $0 = T(v) = T(\alpha_1 v_1 + \cdots + \alpha_n v_n) = \alpha_1 T(v_1) + \cdots + \alpha_n T(v_n) = \alpha_1 w_1 + \cdots + \alpha_n w_n$. Since $\{w_1, \ldots, w_n\}$ is a basis for W it is linearly independent, then $\alpha_1 = \cdots = \alpha_n = 0$ so $v = 0v_1 + \cdots + 0v_n = 0$, so $\ker T \subseteq \{0\}$ and $\{0\} \subseteq \ker T$ it is possible to say $\ker T = \{0\}$, then T is injective.

(b) Let $w \in W$ and $\{w_1, \ldots, w_n\}$ is a basis for W, there are unique $\beta_1, \ldots, \beta_n \in \mathbb{F}$ such that $w = \beta_1 w_1 + \cdots + \beta_n w_n$, then $w = \beta_1 T(v_1) + \cdots + \beta_n T(v_n) = T(\beta_1 v_1 + \cdots + \beta_n v_n) = T(v)$, where$v := \beta_1 v_1 + \cdots + \beta_n v_n \in V$, then T is surjective.

Hence, it is possible to conclude that T is an isomorphism and $V \cong W$. $\square$

7.3 Linear Operators

Definition 7.7 Let V a vector space over a field $\mathbb{F}$. An **linear operator** in V is a linear transformation $T : V \longrightarrow V$.

Proposition 7.8 *Let V a finite-dimensional vector space over a field $\mathbb{F}$ and $T : V \longrightarrow V$ a linear operator. The following statements are equivalent:*

1. *T is injective.*
2. *T is surjective.*
3. *T is bijective.*

Proof 1. Suppose that T is injective, then $\ker T = \{0\}$ and $\dim \ker T = 0$, also V has finite dimension and $\dim \operatorname{Im} T = \dim \ker T + \dim \operatorname{Im} T = \dim V$. Then $\operatorname{Im} T = V$ so T is surjective.

2. Suppose that T is surjective, so $\operatorname{Im} T = V$, since V is of finite dimension, $\dim \ker T = \dim V - \dim \operatorname{Im} T = 0$, then $\ker T = \{0\}$ and T is injective and T is bijective.

3. Clearly, if T is bijective, then T is injective.   □

Corollary 7.1 *Let V and W vector spaces over the field $\mathbb{F}$, with $\dim V = \dim W$. Let $T : V \longrightarrow W$ a linear transformation. The following statements are equivalent:*

1. *T is injective.*
2. *T is surjective.*
3. *T is bijective.*

Proof The proof is similar to the former one, hence it is left as an exercise to the reader.   □

Definition 7.8 Let V and W vector spaces over the field $\mathbb{F}$. The **space of linear transformations** of V in W is given by the following set [5, 12]:

$$\mathscr{L}(V, W) = \{T : V \longrightarrow W \mid T \text{ is a linear transformation}\}$$

For $T_1, T_2, T \in \mathscr{L}(V, W), x \in \mathbb{F}$:

1. $T_1 + T_2 : V \longrightarrow W$ such that $(T_1 + T_2)(\alpha) = T_1(\alpha) + T_2(\alpha)$.
2. $xT : V \longrightarrow W$ such that $(xT)(\alpha) = xT(\alpha)$.

Definition 7.9 Let V a vector space over a field $\mathbb{F}$. The **space of linear operators** of V is given by the following set [3, 4]:

$$\mathscr{L}(V) = \mathscr{L}(V, V) = \{T : V \longrightarrow V \mid T \text{ is a linear operator}\}$$

Proposition 7.9 *Let V and W vector spaces over the field $\mathbb{F}$. If $T_1, T_2 \in \mathscr{L}(V, W)$ and $x_1, x_2 \in \mathbb{F}$ then $x_1 T_1 + x_2 T_2 \in \mathscr{L}(V, W)$.*

Proof Having $x_1T_1 + x_2T_2 : V \to W$ and $v_1, v_2 \in V, a_1, a_2 \in \mathbb{F}$ makes:

$$
\begin{aligned}
(x_1T_1 + x_2T_2)(a_1v_1 + a_2v_2) &= x_1T_1(a_1v_1 + a_2v_2) + x_2T_2(a_1v_1 + a_2v_2) \\
&= x_1(a_1T_1(v_1) + a_2T_1(v_2)) + x_2(a_1T_2(v_1) + a_2T_2(v_2)) \\
&= x_1a_1T_1(v_1) + x_1a_2T_1(v_2) + x_2a_1T_2(v_1) + x_2a_2T_2(v_2) \\
&= a_1[x_1T_1(v_1) + x_2T_2(v_1)] + a_2[x_1T_1(v_2) + x_2T_2(v_2)] \\
&= a_1(x_1T_1 + x_2T_2)(v_1) + a_2(x_1T_1 + x_2T_2)(v_2)
\end{aligned}
$$

Therefore $x_1T_1 + x_2T_2 \in \mathscr{L}(V, W)$ □

Proposition 7.10 *Let V and W vector spaces over the field $\mathbb{F}$ then $\mathscr{L}(V, W)$ is a vector space over $\mathbb{F}$.*

Proof To proof that $L(V, W)$ is a vector space the conditions on scalar multiplication and vector addition must be verified, then it is left to the reader as an exercise. □

Corollary 7.2 *Let V a vector space over $\mathbb{F}$, then $\mathscr{L}(V)$ is a vector space over $\mathbb{F}$.*

Proof It follows from Proposition 7.10 making W=V. □

Proposition 7.11 *Let V and W vector spaces over the field $\mathbb{F}$. If $dimV = n$ and $dimW = m$ then $dim\mathscr{L}(V, W) = mn$.*

Proof Let $\mathbb{A} = \{\alpha_1, \ldots, \alpha_n\}$ and $\mathbb{B} = \{\beta_1, \ldots, \beta_m\}$ be bases for V and W respectively. for every pair of integers (p, q) with $1 \le p \le m$ and $1 \le q \le n$, defining

$$
L^{pq}(\alpha_1) = \begin{cases} 0 & if \ i \ne q \\ \beta_p & if \ i = q \end{cases}
$$

Let T be a linear transformation from V to W, for $1 \le j \le n$, with the elements $A_{ij}, \ldots, A_{mj}$ of $T\alpha_j$ of the basis, this is $T\alpha_j = \sum_{p=1}^{m} A_{pj}\beta_{p^*}$ and it must be shown that $T = \sum_p^m \sum_q^n A_{pq}L^{pq}$ Let

$$
\begin{aligned}
U\alpha_j &= \sum_p \sum_q A_{pq}L^{pq}(\alpha_j) \\
&= \sum_p \sum_q A_{pq}\delta_{jq}\beta_p \\
&= \sum_p^m A_{pj}\beta_p \\
&= T\alpha_j
\end{aligned}
$$

So U=T and L^{pq} spans $L(V, W)$, since they are linearly independent and if the transformation $U = \sum_p \sum_q A_{pq}L^{pq}$ is the zero transformation, then $U\alpha_j = 0$ hence $\sum_{p=1}^{m} A_{pj}\beta_p = 0$ and β_p being linearly independent makes that $A_{pj} = 0$. □

Proposition 7.12 *Let $\{\alpha_1, \alpha_2, \ldots, \alpha_n\}$, $\{\beta_1, \beta_2, \ldots, \beta_m\}$ basis of V and W respectively. A basis for $\mathscr{L}(V, W)$ is*

$$\left\{T_{ij} \mid 1 \le i \le m, 1 \le j \le n\right\}$$

where $T_{ij} : V \longrightarrow W$ is given by $T_{ij}(\alpha_k) = \delta_{jk}\alpha_i$.

Proof It follows from Proposition 7.11 by making $V = W$. $\qquad\square$

Theorem 7.4 *Let V a vector space over $\mathbb{F}$ with $\dim V = n$, then for all $T \in \mathscr{L}(V)$ there exist a polynomial nonzero $q(x) \in \mathbb{F}[x]$ of degree at most n^2 such that $q(T) = 0$.*

Proof Since $\dim \mathscr{L}(V) = n^2$, the set $\left\{Id_V, T, \ldots, T^{n^2}\right\}$ is linearly dependent, then there are $\alpha_0, \alpha_1, \ldots, \alpha_{n^2} \in \mathbb{F}$, not al zero such that $a_0 id_v + a_1 T + \cdots + a_{n^2} T = 0$, the polynomial $q(x))a_0 + a_x x + \cdots + a_{n^2} x^{n^2}$ fulfills the requirement. $\qquad\square$

7.4 Associate Matrix

Let V and W vector spaces over the field $\mathbb{F}$. Let $T : V \longrightarrow W$ a linear transformation. Let $\{\alpha_1, \alpha_2, \ldots, \alpha_n\}$, $\{\beta_1, \beta_2, \ldots, \beta_m\}$ basis of V and W respectively. It is associated with T and bases given, a matrix as follows [2, 8, 11, 12]. There exist $x_{ij} \in \mathbb{F}$ for $1 \le i \le m, 1 \le j \le n$ such that

$$T(\alpha_j) = \sum_{i=1}^{m} x_{ij}\beta_i \text{ for } 1 \le j \le n$$

The matrix

$$A = \left(a_{ij}\right)_{ij} = \begin{bmatrix} a_{11} & a_{12} & \ldots & a_{1n} \\ a_{21} & a_{22} & \ldots & a_{2n} \\ \vdots & & & \vdots \\ a_{m1} & a_{m2} & \ldots & a_{mn} \end{bmatrix}$$

is the **associate matrix** to (or **matrix representation** of) the linear transformation T respect to the basis $\{\alpha_1, \alpha_2, \ldots, \alpha_n\}$, $\{\beta_1, \beta_2, \ldots, \beta_m\}$. To denote this matrix, we write:

$$[T]_{\{\alpha_i\}}^{\{\beta_j\}}$$

Remark 7.4 If T is a linear operator in V and $\{\alpha_1, \alpha_2, \ldots, \alpha_n\}$ is a base of V, the matrix $[T]_{\{\alpha_i\}}^{\{\alpha_i\}}$ is denoted by $[T]_{\{\alpha_i\}}$. If $\alpha \in V$ with $\alpha = x_1\alpha_1 + \cdots + x_n\alpha_n$, the coordinate vector of α respect to the base $\{\alpha_1, \alpha_2, \ldots, \alpha_n\}$ is

$$[\alpha]_{\{\alpha_i\}} = \begin{bmatrix} x_1 \\ \vdots \\ x_n \end{bmatrix}$$

Remark 7.5 If $T \in \mathscr{L}(V, W), \alpha \in V$, then

$$[T]_{\{\alpha_i\}}^{\{\beta_j\}} [\alpha]_{\{\alpha_i\}} = [T(\alpha)]_{\{\beta_i\}}$$

If $T \in \mathscr{L}(V)$, then

$$[T]_{\{\alpha_i\}} [\alpha]_{\{\alpha_i\}} = [T(\alpha)]_{\{\alpha_i\}}$$

Proposition 7.13 *Let V, W, U finite-dimensional vector spaces over a field $\mathbb{F}$. $T, T_1, T_2 : V \longrightarrow W$ and $S : W \longrightarrow U$ linear transformations. Let $\{\alpha_1, \alpha_2, \ldots, \alpha_n\}$, $\{\beta_1, \beta_2, \ldots, \beta_m\}$, $\{\gamma_1, \gamma_2, \ldots, \gamma_p\}$ basis of V, W and U respectively and $x \in \mathbb{F}$, then*

1. $[T_1 + T_2]_{\{\alpha_i\}}^{\{\beta_i\}} = [T_1]_{\{\alpha_i\}}^{\{\beta_i\}} + [T_2]_{\{\alpha_i\}}^{\{\beta_i\}}$.
2. $[xT]_{\{\alpha_i\}}^{\{\beta_i\}} = x [T]_{\{\alpha_i\}}^{\{\beta_i\}}$.
3. $[S \circ T]_{\{\alpha_i\}}^{\{\gamma_i\}} = [S]_{\{\beta_i\}}^{\{\gamma_i\}} [T]_{\{\alpha_i\}}^{\{\beta_i\}}$.
4. *If $n = m$ then T is isomorphism if and only if $[T]_{\{\alpha_i\}}^{\{\beta_i\}}$ is invertible. In addition* $[T^{-1}]_{\{\beta_i\}}^{\{\alpha_i\}} = [T]_{\{\alpha_i\}}^{\{\beta_i\}^{-1}}$.
5. *If $V = W$, then T is isomorphism if and only if $[T]_{\{\alpha_i\}}$ is invertible. In addition* $[T^{-1}]_{\{\alpha_i\}} = [T]_{\{\alpha_i\}}^{-1}$.

Proof Points 3 and 4 will be proven, the remaining are left to the reader as an exercise.

1. $S \circ T : V \to U$, then $[S \circ T]_{\{v_i\}}^{\{u_i\}}$ is a matrix of size $p \times n$ and $[S]_{\{w_i\}}^{\{u_i\}}$ is a matrix $p \times m$ and $[T]_{\{v_i\}}^{\{w_i\}}$ is of $m \times n$, having $[S]_{\{w_i\}}^{\{u_i\}} [T]_{\{v_i\}}^{\{w_i\}}$ be a matrix of size $p \times n$. If $T(v_k) = \sum_{j=1}^{m} b_{jk} w_j$ and $S(w_j) = \sum_{i=1}^{p} a_{ij} u_i$ then:

$$\begin{aligned} (S \circ T)(v_k) &= S(T(v_k)) \\ &= S\left(\sum_{j=1}^{m} b_{jk} w_j\right) \\ &= \sum_{j=1}^{m} b_{jk} S(w_j) \\ &= \sum_{j=1}^{m} b_{jk} \sum_{i=1}^{p} a_{ij} u_i \\ &= \sum_{i=1}^{p} \left(\sum_{j=1}^{m} a_{ij} b_{jk}\right) u_i \end{aligned}$$

The ij entry of $S^{\{u_i\}}_{\{w_i\}}$ is a_{ij}, the entry jk of $[T]^{\{w_i\}}_{\{v_i\}}$ is b_{ij} and the entry ik of $[S]^{\{u_i\}}_{\{w_i\}}[T]^{\{w_i\}}_{\{v_i\}}$ is $\sum_{j=1}^m a_{ij}b_{jk}$, also the value ik of $[S \circ T]^{\{u_i\}}_{\{v_i\}}$ is $\sum_{j=1}^m a_{ij}b_{jk}$, hence
$$[S \circ T]^{\{u_i\}}_{\{v_i\}} = [S]^{\{u_i\}}_{\{w_i\}}[T]^{\{w_i\}}_{\{v_i\}}$$

2. Suppose that $[T]^{\{w_i\}}_{\{v_i\}}$ is invertible, there is a matrix $B = (b_{jk})$ such that $[T]^{\{w_i\}}_{\{v_i\}} B = I_n$, Making $S : W \to V$ for $S(w_j) = \sum_{i=1}^n b_{ij}v_i$ with the base $\{w_1, \ldots, w_m\}$, then S is a linear transformation and $[S]^{\{u_i\}}_{\{w_i\}} = B$, let $w \in W$, then
$$T(S(w))_{\{wi\}} = (T \circ S)_{\{wi\}}[w]_{\{wi\}} = [T]^{\{w_i\}}_{\{v_i\}} S^{\{ui\}}_{\{wi\}}[w]_{\{wi\}} = I_n[w]_{\{wi\}} = [w]_{\{wi\}},$$
then $T(S(w)) = w$ allows to conclude that T is surjective, and being V and W of the dame finite dimension and T a linear transformation, it is bijective and a isomorphism.   □

Theorem 7.5 *Let* V, W *finite-dimensional vector spaces over a field* $\mathbb{F}$ *and* $\{\alpha_1, \alpha_2, \ldots, \alpha_n\}$, $\{\beta_1, \beta_2, \ldots, \beta_m\}$ *basis of* V, W *respectively. The function*

$$\Phi : \mathscr{L}(V, W) \longrightarrow \mathscr{M}_{m \times n}(\mathbb{F})$$

with $T \longmapsto [T]^{\{\beta_i\}}_{\{\alpha_i\}}$ *is an isomorphism of vector spaces.*

Proof The function is linear and a one on one map $L(V; W)$ into the set of matrices $m \times n$.   □

Corollary 7.3 *Let* V *a finite-dimensional vector space over a field* $\mathbb{F}$ *and* $\{\alpha_1, \alpha_2, \ldots, \alpha_n\}$ *a basis of* V. *The function*

$$\Phi : \mathscr{L}(V) \longrightarrow \mathscr{M}_{n \times n}(\mathbb{F})$$

with $T \longmapsto [T]_{\{\alpha_i\}}$ *is an isomorphism of vector spaces.*

Proof It follows from the theorem by making $V = W$.   □

Remark 7.6 The isomorphism in **Corollary** 7.3 is called an **isomorphism of algebras** becaus $\mathscr{L}(V)$ and $\mathscr{M}_{n \times n}(\mathbb{F})$ have the following properties:

1. $\mathscr{L}(V)$ has the composition of linear operators.
2. $\mathscr{M}_{n \times n}(\mathbb{F})$ has the product of square matrix.

These properties are inherited at ϕ function, i.e.:

$$\Phi(S \circ T) = [S \circ T]_{\{\alpha_i\}} = [S]_{\{\alpha_i\}} \cdot [T]_{\{\alpha_i\}} = \Phi(S) \cdot \Phi(T)$$

Theorem 7.6 *Let* V, W *finite-dimensional vector spaces over a field* $\mathbb{F}$. *Let* $\{\alpha_1, \alpha_2, \ldots, \alpha_n\}$ *and* $\{\alpha'_1, \alpha'_2, \ldots, \alpha'_n\}$ *basis of* V. *Respectively, let* $\{\beta_1, \beta_2, \ldots, \beta_m\}$, $\{\beta'_1, \beta'_2, \ldots, \beta'_m\}$ *basis of* W *and* $T \in \mathscr{L}(V, W)$. *Let* P *the change of basis matrix of basis* $\{\alpha_1, \alpha_2, \ldots, \alpha_n\}$ *to basis* $\{\alpha'_1, \alpha'_2, \ldots, \alpha'_n\}$ *of* V *and* Q *the change of basis matrix of basis* $\{\beta_1, \beta_2, \ldots, \beta_m\}$ *to basis* $\{\beta'_1, \beta'_2, \ldots, \beta'_m\}$ *of* W. *We have*

$$[T]_{\{\alpha'_i\}}^{\{\beta'_i\}} = Q^{-1}[T]_{\{\alpha_i\}}^{\{\beta_i\}}P$$

Proof It is left to the reader as an exercise. □

Corollary 7.4 *Let* V *a finite-dimensional vector space over a field* $\mathbb{F}$. *Let* $\{\alpha_1, \alpha_2, \ldots, \alpha_n\}$ *and* $\{\alpha'_1, \alpha'_2, \ldots, \alpha'_n\}$ *basis of* V *and* $T \in \mathcal{L}(V)$. *Let* P *the change of basis matrix of basis* $\{\alpha_1, \alpha_2, \ldots, \alpha_n\}$ *to basis* $\{\alpha'_1, \alpha'_2, \ldots, \alpha'_n\}$ *of* V, *then:*

$$[T]_{\{\alpha'_i\}} = P^{-1}[T]_{\{\alpha_i\}}P$$

Proof Consider $v \in V$ and knowing that $P[v]_{\{\alpha'_i\}} = [v]_{\{\alpha_i\}}$ makes $P^{-1}[T]_{\{\alpha_i\}}$ $P[v]_{\{\alpha'_i\}} = P^{-1}[T]_{\{\alpha_i\}}[v]_{\{\alpha_i\}} = P^{-1}[T(v)]_{\{\alpha_i\}} = [T(v)]_{\{\alpha'_i\}}$ but $[T]_{\{\alpha'_i\}}[v]_{\{\alpha'_i\}} = [T(v)]_{\{\alpha'_i\}}$, then $P^{-1}[T]_{\{\alpha_i\}}P[v]_{\{\alpha'_i\}} = [T]_{\{\alpha'_i\}}[v]_{\{\alpha'_i\}}$.

Since $v \to [v]_{\{\alpha'_i\}}$ is in $\mathbb{F}^n$, it leads to $P^{-1}[T]_{\{\alpha_i\}}PX = [T]_{\{\alpha'_i\}}X$ for any $X \in \mathbb{F}^n$, therefore $P^{-1}[T]_{\{\alpha_i\}}P = [T]_{\{\alpha'_i\}}$ □

Lemma 7.1 *Let* $P = (c_{hi})_{h,i} \in \mathcal{M}_{n \times n}$, *an invertible matrix and* $\{v_1, v_2, \ldots, v_n\}$ *a linearly independent set. For* $1 \leq i \leq n$, *let*

$$v'_i = \sum_{h=1}^{n} c_{hi}v_h$$

Then $\{v'_1, v'_2, \ldots, v'_n\}$ *is linearly independent.*

Proof Consider $x_1v'_1 + \cdots + x_nv'_n = 0$ for some $x_1, \ldots, x_n \in \mathbb{R}$, having:

$$0 = \sum_{i=1}^{n} x_iv'_i$$

$$= \sum_{i=1}^{n} x_i \sum_{h=1}^{n} c_{hi}v_h$$

$$= \sum_{h=1}^{n} \left(\sum_{i=1}^{n} x_ic_{hi}\right)v_h$$

The set $\{v_1, \ldots, v_n\}$ is linearly independent and $\sum_{i=1}^{n} x_ic_{hi} = 0$ for $1 \leq h \leq n$ make the system:

$$c_{11}x_1 + \cdots + x_{1n}x_n = 0$$
$$c_{21}x_1 + \cdots + x_{2n}x_n = 0$$
$$\vdots \; 0$$
$$c_{n1}x_1 + \cdots + x_{nn}x_n = 0$$

The associated matrix for the systems is P, making the proper elementary operations P can be made into the identity, and the systems will be:

$$x_1 = 0$$
$$x_2 = 0$$
$$\vdots$$
$$x_n = 0$$

Then $\{v_i', \ldots, v_n'\}$ is linearly independent. $\square$

Definition 7.10 Let V a vector space over $\mathbb{F}$, with dim $V = n$. Let $T : V \longrightarrow V$ a linear operator and $A \in \mathcal{M}_{n \times n}(\mathbb{F})$. It said to be matrix A **represents to** T if there exists a basis $\{\alpha_1, \alpha_2, \ldots, \alpha_n\}$ of V such that [9, 10, 16]:

$$[T]_{\{v_i\}} = A$$

Theorem 7.7 *Let $A, B \in \mathcal{M}_{n \times n}(\mathbb{F})$ and V a vector space over $\mathbb{F}$, with dim $V = n$. The matrices A and B represent the same linear operator T in V if and only if A and B are similar.*

Proof There are $T \in \mathcal{L}(V)$ and $\{v_1, \ldots, v_n\}$, $\{v_1', \ldots, v_n'\}$ are bases for V, then $A = [T]_{\{v_i\}}$ and $B = [T]_{\{v_i'\}}$. Let P be a matrix for basis change from $\{v_i, \ldots, v_n\}$ to $\{v_i', \ldots, v_n'\}$, then $B = P^{-1}AP$ implies that A and B are similar.

Similarly, Suppose that $A = (a_{ij})$ and B are similar, then, there is $P \in \mathcal{M}_{m \times n}(\mathbb{F})$ invertible such that $B = P^{-1}AP$, let $\{v_i, \ldots, v_n\}$ be a basis for V. Making $T : V \to V$ in the basis $\{v_1, \ldots, v_n\}$ for $T(v_j) = \sum_{i=1}^{n} a_{ij}v_i$ for $1 \leq j \leq n$ and completing for linearity, then $T \in \mathcal{L}(V)$ and A represent the operator T corresponding to the basis $\{v_i, \ldots, v_n\}$. Also if $P = (c_{hi})_{hi}$ for $1 \leq i \leq n$, $v_i' = \sum_{h=1}^{n} c_{hi}v_h$, the set $\{v_1', \ldots, v_n'\}$ is linearly independent, then it is also a basis for V having $B = P^{-1}AP = P^{-1}[T]_{\{v_i\}} P = [T]_{\{v_i'\}}$, hence B also represents T. $\square$

References

1. Hoffman, K., Kunze, R.: Linear Algebra. Prentice-Hall (1971)
2. Grossman, S.: Elementary Linear Algebra. Wadsworth Publishing (1987)
3. Whitelaw, T.: Introduction to Linear Algebra. Chapman & Hall/CRC (1992)
4. Anton, H.: Elementary Linear Algebra. Wiley (2010)
5. Gantmacher, F.R.: The Theory of Matrices, vol. 2. Chelsea, New York (1964)
6. Axler, S.: Linear Algebra Done-Right. Springer (1997)
7. Lipschutz, S.: Linear Algebra. McGraw Hill (2009)
8. Nef, W.: Linear Algebra. Dover Publications (1967)
9. Lang, S.: Linear Algebra. Springer (2004)
10. Nering, E.: Linear Algebra and Matrix Theory. Wiley (1970)

11. kolman, B.: Introductory Linear Algebra an Applied First Course. Pearson Education (2008)
12. Gantmacher, F.R.: The Theory of Matrices, vol. 1. Chelsea, New York (1959)
13. Kailath, T.: Linear Systems. Prentice-Hall, Englewood Cliffs, NJ (1980)
14. Hermann, R., Krener, A.: Nonlinear controllability and observability. IEEE Trans. Autom. Control **22**(5), 728–740 (1977)
15. Aguilar-Lpez, R., Mata-Machuca, J.L., Martnez-Guerra, R.: Observability and Observers for Nonlinear Dynamical Systems: Nonlinear Systems Analysis. Lambert Academic Publishing (2011)
16. Halmos, P. R.: Finite-Dimensional Vector Spaces. Courier Dover Publications (2017)

Chapter 8
Matrix Diagonalization and Jordan Canonical Form

Abstract This chapter focuses on the basic theory of Matrix Diagonalization and Jordan Canonical Form.

8.1 Matrix Diagonalization

Let V a finite dimensional vector space, dim $V = n$ and $T : V \to V$ a linear operator, then T can be represented by a matrix $A \in \mathcal{M}_{n \times n}(\mathbb{R})$. For this reason, some occasions we are going to refer to eigenvalues and eigenvectors of $n \times n$.

Theorem 8.1 *Let $A \in \mathcal{M}_{n \times n}(\mathbb{R})$, then λ is a eigenvalue of A if and only if*

$$p(\lambda) = \det(A - \lambda I) = 0$$

Proof The number λ is an eigenvalue of A if and only if there is a non zero vector v that fulfills:

$$Av = \lambda v$$
$$\lambda I v - Av = 0$$
$$(\lambda I - A) v = 0$$

If not the matrix $(\lambda I - A)$ is singular. The value λ is a root of $P(\lambda) = |\lambda I - A|$, then v is an Eigenvector only if the statement is true, hence it is a solution of the characteristic polynomial □

We know that $p(\lambda)$ can be written as:

$$p(\lambda) = \lambda^n + a_{n-1}\lambda^{n-1} + \cdots + a_1\lambda + a_0 = 0$$

© Springer Nature Switzerland AG 2019
R. Martínez-Guerra et al., *Algebraic and Differential Methods for Nonlinear Control Theory*, Mathematical and Analytical Techniques with Applications to Engineering, https://doi.org/10.1007/978-3-030-12025-2_8

This equation has n roots, which can be repeated. If $\lambda_1, \lambda_2, \ldots, \lambda_k$ are the different roots of $p(\lambda)$ with multiplicities $r_1, r_2, \ldots, r_k$ respectively, then $p(\lambda)$ can be factorized as:

$$p(\lambda) = (\lambda - \lambda_1)^{r_1} (\lambda - \lambda_2)^{r_2} \cdots (\lambda - \lambda_k)^{r_k} = 0.$$

Definition 8.1 It said that the numbers $r_1, r_2, \ldots, r_k$ are called **algebraic multiplicities** of eigenvalues $\lambda_1, \lambda_2, \ldots, \lambda_k$ respectively [1–4].

Definition 8.2 Let λ an eigenvalue of T and

$$E_\lambda = \{v \mid Tv = \lambda v\} = \{v \mid (T - \lambda I)v = 0\}$$

The subspace E_λ (since is the kernel of linear transformation $T - \lambda I$) is called **Eigenspace** of T corresponding to eigenvalue λ.

Since E_λ is a subspace then $0 \in E_\lambda$, but the dimension dim $E_\lambda > 0$ since by definition if λ is an eigenvalue then there exists an eigenvalue nonzero corresponding to λ.

Theorem 8.2 *Let V a finite dimensional vector space, $\dim V = n$ and $T : V \to V$ a linear operator. Let $\lambda_1, \lambda_2, \ldots, \lambda_k$ different eigenvalues of T and their corresponding eigenvectors $v_1, v_2, \ldots, v_n$, then $v_1, v_2, \ldots, v_n$ are linearly independent.*

Proof Suppose that $v_1, \ldots, v_n$ are the minimal set of vectors for which the theorem is not true, having $s > 1$ and $v_1 \neq 0$, by the minimality condition $v_2, \ldots, v_s$ are linearly independent, then v_1 is a linear combination of $v_2, \ldots, v_s$:

$$v_1 = a_2 v_2 + \cdots + a_s v_s$$

Applying T makes:

$$T(v_1) = T(a_2 v_2 + \cdots + a_s v_s)$$
$$= T(a_2 v_2) + \cdots + T(a_s v_s)$$

Consider the eigenvector v_j corresponding to the eigenvalue λ_j and applying T $T(v_j) = \lambda_j v_j$, then:

$$\lambda_1 v_1 = a_2 \lambda_2 v_2 + \cdots + a_s \lambda_s v_s$$

Multiplying by λ_1:

$$\lambda_1 v_1 = a_2 \lambda_1 v_2 + \cdots + a_s \lambda_1 v_s$$

$$a_2 (\lambda_1 - \lambda_2) v_2 + \cdots + a_s (\lambda_1 - \lambda_s) v_s = 0$$

Being $v_2, \ldots, v_s$ linearly independent causes that $a_2 (\lambda_1 - \lambda_2) = 0, \ldots, a_s$ $(\lambda_1 - \lambda_s) = 0$, but the values of λ are different, then $\lambda_1 \neq \lambda_j, \quad j > 1$, hence $a_2 = 0, \ldots, a_s = 0$ contradicting $a_k \neq 0$. $\qquad\square$

Remark 8.1 Let A a matrix representation of T. Let suppose $A \in \mathcal{M}_{3\times 3}(\mathbb{R})$ with eigenvalues λ_1, λ_2 and λ_3 and v_1, v_2 and v_3 the eigenvectors associated with each λ [5–8]:

(i) If the algebraic multiplicity of each λ is 1, by Theorem 8.1, v_1, v_2 and v_3 are linearly independent.
(ii) If the algebraic multiplicity of λ_1 and λ_2 is 1 and 2 respectively, then dim $E_{\lambda_2} \leq$ 2. Otherwise we could have at least 4 linearly independent vectors in a space of three dimensions, that is, for λ_2 there can be 1 or 2 linearly independent eigenvectors.
(iii) If the algebraic multiplicity of λ_1 is 3 then dim $E_\lambda \leq 3$. This is, there can be 1, 2 or 3 linearly independent eigenvectors.

Theorem 8.3 *Let V a finite dimensional vector space,* dim $V = n$ *and $T : V \to V$ a linear operator. Let λ an eigenvalue of T, then:*

$$1 \leq \dim E_\lambda \leq \text{algebraic multiplicity of } \lambda.$$

Proof It is left to the reader as an exercise. $\qquad\square$

Example 8.1 Let

$$A = \begin{bmatrix} -3 & 3 & -2 \\ -7 & 6 & -3 \\ 1 & -1 & 2 \end{bmatrix}$$

The eigenvalues of A corresponding to the roots of $p(\lambda)$, this is:

$$p(\lambda) = \det (A - \lambda I) = (\lambda - 2)^2 (\lambda - 1) = 0$$

hence, the eigenvalues are:

- $\lambda_1 = 2$ with algebraic multiplicity 2.
- $\lambda_2 = 1$ with algebraic multiplicity 1.

On the other hand:

$$E_{\lambda_1} = \ker(A - \lambda_1 I)$$
$$= \ker(A - 2I)$$

$$E_{\lambda_1} = \left\{ \begin{bmatrix} x_1 \\ x_2 \\ x_3 \end{bmatrix} : \begin{bmatrix} -5 & 3 & -2 \\ -7 & 4 & -3 \\ 1 & -1 & 0 \end{bmatrix} \begin{bmatrix} x_1 \\ x_2 \\ x_3 \end{bmatrix} = \begin{bmatrix} 0 \\ 0 \\ 0 \end{bmatrix} \right\}$$

The solution of above system of linear equations provides the unique linearly independent eigenvector given by:

$$v_1 = \begin{bmatrix} 1 \\ 1 \\ -1 \end{bmatrix}$$

then $E_{\lambda_1} = \mathscr{L}\{v_1\}$, hence dim $E_{\lambda_1} = 1$. Similarly for λ_2:

$$E_{\lambda_2} = \ker(A - \lambda_2 I)$$
$$= \ker(A - I)$$

$$E_{\lambda_2} = \left\{ \begin{bmatrix} x_1 \\ x_2 \\ x_3 \end{bmatrix} : \begin{bmatrix} -4 & 3 & -2 \\ -7 & 5 & -3 \\ 1 & -1 & 1 \end{bmatrix} \begin{bmatrix} x_1 \\ x_2 \\ x_3 \end{bmatrix} = \begin{bmatrix} 0 \\ 0 \\ 0 \end{bmatrix} \right\}$$

Solving the above system of linear equations, the linearly independent eigenvector associated to λ_2 is

$$v_2 = \begin{bmatrix} 1 \\ 2 \\ 1 \end{bmatrix}$$

then $E_{\lambda_2} = \mathscr{L}\{v_2\}$, hence dim $E_{\lambda_2} = 1$.

Definition 8.3 Let $A, B \in \mathscr{M}_{n \times n}(\mathbb{R})$. It said A and B are **equivalent matrices** if there exists an invertible matrix $Q \in \mathscr{M}_{n \times n}(\mathbb{R})$ such that:

$$B = Q^{-1}AQ$$

In addition, Q is called **transformation matrix**.

Theorem 8.4 *If $A, B \in \mathscr{M}_{n \times n}(\mathbb{R})$ are equivalent matrices then A and B have the same characteristic polynomial and therefore they have the same eigenvalues.*

Proof Having the invertible matrix P that fulfills $B = P^{-1}AP$, which also makes $\lambda I = P^{-1}\lambda I P$ such that:

$$|\lambda I - B| = |\lambda I - P^{-1}AP|$$
$$= |P^{-1}\lambda I P - P^{-1}AP|$$
$$= |P^{-1}(\lambda I - A)P|$$
$$= |P^{-1}||\lambda I - A||P|$$

Knowing that $|P^{-1}||P| = 1$, then $|\lambda I - B| = |\lambda I - A|$. □

Definition 8.4 Let $A \in \mathcal{M}_{n \times n}(\mathbb{R})$. It said that A is a **diagonalizable matrix** if there exist a diagonal matrix D such that A and D are equivalent, that is to say:

$$D = Q^{-1}AQ$$

where $Q \in \mathcal{M}_{n \times n}(\mathbb{R})$ is a invertible matrix.

Remark 8.2 • If D is a diagonal matrix, the eigenvalues of D are the components of the principal diagonal.
• If A is equivalent to D, by Theorem 8.4, the eigenvalues of A are the same that D. According to above comments, if A is diagonalizable then A is equivalent to a diagonal matrix with components in the principal diagonal are the eigenvalues of A. In addition, given a linear transformation $T : V \to V$ in a vector space V with $\dim V = n$, it is worth asking:

1. Is there a base β for V such that the matrix representation A of T is a diagnonalizable matrix?
2. If that base exists, how can it be found?

The following theorem has the answer.

Theorem 8.5 *Let $A \in \mathcal{M}_{n \times n}(\mathbb{R})$. A is diagonalizable if and only if A has n linearly independent eigenvectors. In this case, the diagonal matrix D equivalent to A is given by:*

$$D = \begin{bmatrix} \lambda_1 & 0 & 0 & \cdots & 0 \\ 0 & \lambda_2 & 0 & \cdots & 0 \\ 0 & 0 & \lambda_3 & \cdots & 0 \\ \vdots & \vdots & \vdots & \ddots & \vdots \\ 0 & 0 & 0 & \cdots & \lambda_n \end{bmatrix}$$

where $\lambda_1, \lambda_2, \ldots, \lambda_n$ are the eigenvalues of A. In addition if Q is a matrix whose columns are linearly independent vectors of A, then:

$$D = Q^{-1}AQ.$$

Proof The matrix A has eigenvalues $\lambda_1, \ldots, \lambda_n$ that make linearly independent eigenvectors $v_1, \ldots, v_n$, let the Q matrix have in its columns $v_1, \ldots, v_n$, so it is non singular. The columns of AQ are $Av_1, \ldots, Av_n$ and $Av_k = \lambda v_k$, then its columns can be written as $\lambda_1 v_1, \ldots, \lambda_n v_n$, the matrix $D = diag\,(\lambda_1, \ldots, \lambda_n)$ implies that QD has columns $\lambda_k v_k$, then

$$AQ = QD$$

And

$$D = Q^{-1}AQ$$

Suppose there is a non singular matrix Q such that $Q^{-1}AQ = diag\,(\lambda_1, \ldots, \lambda_n) = D$ and $v_1, \ldots, v_n$ are the columns of Q, then the columns of $AQ = Av_k$ and the columns of $QD = \lambda_k v_k$ then $Av_1 = \lambda_1 v_1, \cdots, Av_n = \lambda_n v_n$, even more, Q is non-singular and $v_1, \ldots, v_n$ are non zero, eigenvectors of A, diagonal elements of D and linearly independent, hence the theorem is true. $\qquad\square$

Corollary 8.1 *If $A \in \mathcal{M}_{n \times n}(\mathbb{R})$ has n different eigenvalues then A is diagonalizable.*

Proof It is left to the reader as an exercise. $\qquad\square$

Example 8.2 Let

$$A = \begin{bmatrix} 1 & -1 & 4 \\ 3 & 2 & -1 \\ 2 & 1 & -1 \end{bmatrix}$$

with eigenvalues given by $p(\lambda) = \det(A - \lambda I) = -(\lambda - 1)(\lambda + 2)(\lambda - 3) = 0$, hence $\lambda_1 = 1, \lambda_2 = -2, \lambda_3 = 3$. Since the eigenvalues are different, by Theorem 8.2 there are three linearly independent eigenvectors that can be:

$$v_1 = \begin{bmatrix} -1 \\ 4 \\ 1 \end{bmatrix}, v_2 = \begin{bmatrix} 1 \\ -1 \\ -1 \end{bmatrix}, v_3 = \begin{bmatrix} 1 \\ 2 \\ 1 \end{bmatrix}$$

Hence, by Theorem 8.5 A is diagonalizable where:

$$Q = \begin{bmatrix} -1 & 1 & 1 \\ 4 & -1 & 2 \\ 1 & -1 & 1 \end{bmatrix}$$

Finally:

$$D = Q^{-1}AQ = -\frac{1}{6} \begin{bmatrix} 1 & -2 & 3 \\ -2 & -2 & 6 \\ -3 & 0 & -3 \end{bmatrix} \begin{bmatrix} 1 & -1 & 4 \\ 3 & 2 & -1 \\ 2 & 1 & -1 \end{bmatrix} \begin{bmatrix} -1 & 1 & 1 \\ 4 & -1 & 2 \\ 1 & -1 & 1 \end{bmatrix} = \begin{bmatrix} 1 & 0 & 0 \\ 0 & -2 & 0 \\ 0 & 0 & 3 \end{bmatrix}$$

Example 8.3 Although the eigenvalues of a matrix are not different, it is possible to diagonalize the given matrix. Let

$$A = \begin{bmatrix} 3 & 2 & 4 \\ 2 & 0 & 2 \\ 4 & 2 & 3 \end{bmatrix}$$

with eigenvalues given by $p(\lambda) = \det(A - \lambda I) = -(\lambda + 1)^2(\lambda - 8) = 0$, hence $\lambda_1 = -1$ with algebraic multiplicity 2 and $\lambda_2 = 8$. For $\lambda_2 = 8$ the linearly independent eigenvector is:

$$v_1 = \begin{bmatrix} 2 \\ 1 \\ 2 \end{bmatrix}$$

On the other hand for $\lambda_1 = -1$ the corresponding linearly independent eigenvectors are:

$$v_2 = \begin{bmatrix} 1 \\ -2 \\ 0 \end{bmatrix}, v_3 = \begin{bmatrix} 0 \\ -2 \\ 1 \end{bmatrix}$$

By Theorem 8.5 the matrix A is diagonalizable with

$$Q = \begin{bmatrix} 2 & 1 & 0 \\ 1 & -2 & -2 \\ 2 & 0 & 1 \end{bmatrix}$$

$$D = Q^{-1}AQ = -\frac{1}{9} \begin{bmatrix} -2 & -1 & -2 \\ -5 & 2 & 4 \\ 4 & 2 & -5 \end{bmatrix} \begin{bmatrix} 3 & 2 & 4 \\ 2 & 0 & 2 \\ 4 & 2 & 3 \end{bmatrix} \begin{bmatrix} 2 & 1 & 0 \\ 1 & -2 & -2 \\ 2 & 0 & 1 \end{bmatrix} = \begin{bmatrix} 8 & 0 & 0 \\ 0 & -1 & 0 \\ 0 & 0 & -1 \end{bmatrix}$$

Example 8.4 In Example 8.1, the matrix A has only two linearly independent eigenvectors, then it is not possible to find the transformation matrix Q. Hence by Theorem 8.5, A is not diagonalizable.

The matrices $M \in \mathcal{M}_{n \times n}(\mathbb{R})$ with n linearly independent eigenvectors can be expressed in a diagonal matrix through a equivalence transformation. Most polynomials have different roots, then most of the matrices will have different eigenvalues and therefore they will be diagonalizable [3–5, 8].

Matrices that are not diagonalizable (matrices that do not have n linearly independent eigenvectors) have some applications. In this case, it still possible to prove that matrix is equivalent to other matrix most simple but the new matrix is not diagonal and the transformation matrix Q is more difficult to obtain.

8.2   Jordan Canonical Form

Although not every linear operator T is diagonalizable, it is possible to find a base β for the vector space V with $\dim V = n$ such that the matrix representation $A \in \mathcal{M}_{n \times n}(\mathbb{R})$ of T is equivalent to:

$$J = \begin{bmatrix} J_1 & 0 & 0 & \cdots & 0 \\ 0 & J_2 & 0 & \cdots & 0 \\ 0 & 0 & J_3 & \cdots & 0 \\ \vdots & \vdots & \vdots & \ddots & \vdots \\ 0 & 0 & 0 & \ldots & J_n \end{bmatrix}$$

where J_i is a matrix of the form (λ_i) or the form:

$$\begin{bmatrix} \lambda_i & 1 & 0 & 0 & \cdots & 0 \\ 0 & \lambda_i & 1 & 0 & \cdots & 0 \\ 0 & 0 & \lambda_i & 1 & \cdots & 0 \\ \vdots & \vdots & \vdots & \ddots & \ddots & \vdots \\ 0 & 0 & 0 & \ldots & \lambda_i & 1 \\ 0 & 0 & 0 & \ldots & 0 & \lambda_i \end{bmatrix}$$

for some eigenvalue λ_i of A. In the above description:

- J_i is called **Jordan block** corresponding to λ_i.
- β_i is the **corresponding base** to the block J_i.
- J is called **Jordan canonical form** of A.
- β is the **Jordan canonical base**.

Remark 8.3 Each Jordan block J_i is almost a diagonal matrix. J is a diagonal matrix if and only if each J_i is of the form (λ_i) [9–11].

Example 8.5

$$\begin{bmatrix} 4 & \vdots & 0 & 0 & 0 & \vdots & 0 \\ \cdots & & \cdots & \cdots & \cdots & & \cdots \\ 0 & \vdots & -3 & 1 & 0 & \vdots & 0 \\ 0 & \vdots & 0 & -3 & 1 & \vdots & 0 \\ 0 & \vdots & 0 & 0 & -3 & \vdots & 0 \\ \cdots & & \cdots & \cdots & \cdots & & \cdots \\ 0 & \vdots & 0 & 0 & 0 & \vdots & 7 \end{bmatrix}$$

Jordan blocks are marked with dotted lines. The algebraic multiplicity of each eigenvalue λ_i is the number of times that the eigenvalue λ_i appears on the diagonal of J_i.

Proposition 8.1 *Let λ_i an eigenvalue of A such that $\dim E_{\lambda_i} = s_i$, then the number of ones above the diagonal of Jordan canonical form is given by:*

$$n - \sum_{i=1}^{k} s_i \qquad (8.1)$$

where k is the number of eigenvalues.

Proof It is left to the reader as an exercise. $\qquad\qquad\qquad\qquad\qquad\qquad$ ☐

Example 8.6 If the characteristic polynomial of a matrix $A \in \mathcal{M}_{4 \times 4}(\mathbb{R})$ is

$$p(\lambda) = (\lambda - 2)^3 (\lambda + 3)$$

then $\lambda_1 = 2$ with algebraic multiplicity 3 and $\dim E_{\lambda_1} \leq 3$. For $\lambda_2 = -3$ with algebraic multiplicity 1 and $\dim E_{\lambda_2} = 1$. Since $\dim E_{\lambda_1} \leq 3$ then there can be 1, 2 or 3 linearly independent eigenvectors for λ_1. For λ_2 there is always a linearly independent eigenvector, hence the possible Jordan canonical forms of A are:

1. If $\dim E_{\lambda_1} = 1$, by Eq. (8.1) the number of ones above the diagonal of Jordan canonical form is $3 - 1 = 2$:

$$J_1 = \begin{bmatrix} 2 & 1 & 0 & 0 \\ 0 & 2 & 1 & 0 \\ 0 & 0 & 2 & 0 \\ 0 & 0 & 0 & -3 \end{bmatrix}$$

2. If $\dim E_{\lambda_1} = 2$, the number of ones above the diagonal of Jordan canonical form is $3 - 2 = 1$:

$$J_2 = \begin{bmatrix} 2 & 1 & 0 & 0 \\ 0 & 2 & 0 & 0 \\ 0 & 0 & 2 & 0 \\ 0 & 0 & 0 & -3 \end{bmatrix}$$

3. If $\dim E_{\lambda_1} = 3$, in this case the number of ones above the diagonal of Jordan canonical form is $3 - 3 = 0$:

$$J_3 = \begin{bmatrix} 2 & 0 & 0 & 0 \\ 0 & 2 & 0 & 0 \\ 0 & 0 & 2 & 0 \\ 0 & 0 & 0 & -3 \end{bmatrix}$$

In the columns where appear the ones above the eigenvalue indicate that there are vectors that are not eigenvectors. In J_1 the columns 2 and 3 indicate that for λ_1 there are two vectors that are not eigenvectors. On the other hand for J_3 all vectors are eigenvectors.

Theorem 8.6 *Let $A \in \mathcal{M}_{n \times n}(\mathbb{R})$, then there exists an invertible matrix $Q \in \mathcal{M}_{n \times n}$ ($\mathbb{R}$) such that:*

$$J = Q^{-1} A Q$$

where J is a Jordan matrix whose diagonal elements are the eigenvalues of A. In addition, J is unique except by the order of the Jordan blocks.

Proof It is left to the reader as an exercise.   □

Example 8.7 From Example 8.6, if A is equivalent to J_1 then A is equivalent to:

$$\begin{bmatrix} -3 & 0 & 0 & 0 \\ 0 & 2 & 1 & 0 \\ 0 & 0 & 2 & 1 \\ 0 & 0 & 0 & 2 \end{bmatrix}$$

this is, the Jordan blocks are the same but the order can be different.

Example 8.8 According to the methodology to find the matrix representation A of a linear transformation T, let $P_2(\mathbb{R})$ the set of polynomials and $\beta = \{x_1, x_2, x_3\} = \{1, x, x^2\}$ a base for $P_2(\mathbb{R})$. Let define $T : P_2(\mathbb{R}) \rightarrow P_2(\mathbb{R})$ through $T(f) = f'$. The matrix representation A can be calculated as follows:

$$\begin{aligned} T(1) &= 0 = 0 \cdot 1 + 0 \cdot x + 0 \cdot x^2 \\ T(x) &= 1 = 1 \cdot 1 + 0 \cdot x + 0 \cdot x^2 \\ T(x^2) &= 2x = 0 \cdot 1 + 2 \cdot x + 0 \cdot x^2 \end{aligned}$$

then

$$A = \begin{bmatrix} 0 & 1 & 0 \\ 0 & 0 & 2 \\ 0 & 0 & 0 \end{bmatrix}.$$

8.2.1   Generalized Eigenvectors

Definition 8.5 Let T a linear operator in a space vector V with dim $V = n$. An element nonzero $v \in V$ is called **generalized eigenvector** of T if there exists an scalar λ such that $(T - \lambda I)^p (v) = 0$ for any $p \in \mathbb{Z}^+$.

Definition 8.6 It said to be v is a generalized eigenvector corresponding to λ.

Example 8.9 Let

$$J = \begin{bmatrix} 2 & 1 & 0 & 0 & 0 & 0 & 0 & 0 \\ 0 & 2 & 1 & 0 & 0 & 0 & 0 & 0 \\ 0 & 0 & 2 & 0 & 0 & 0 & 0 & 0 \\ 0 & 0 & 0 & 2 & 0 & 0 & 0 & 0 \\ 0 & 0 & 0 & 0 & 3 & 1 & 0 & 0 \\ 0 & 0 & 0 & 0 & 0 & 3 & 0 & 0 \\ 0 & 0 & 0 & 0 & 0 & 0 & 0 & 1 \\ 0 & 0 & 0 & 0 & 0 & 0 & 0 & 0 \end{bmatrix}$$

The characteristic polynomial is $p(\lambda) = \det(J - \lambda I) = (\lambda - 2)^4 (\lambda - 3)^2 \lambda^2$ where $\lambda_1 = 2$ with algebraic multiplicity 4, $\lambda_2 = 3$ with algebraic multiplicity 2 and $\lambda_3 = 0$ with algebraic multiplicity 2. From the associated basic vectors $x_1, x_2, \ldots, x_8$ to λ_1, λ_2 and λ_3 only x_1, x_4, x_5 and x_7 are eigenvectors. Since x_1, x_4 are corresponding eigenvectors to λ_1 then:

$$(T - 2I)(x_1) = (T - 2I)(x_4) = 0$$

but x_2 and x_3 are not eigenvectors. Hence:

$$(T - 2I)(x_2) \neq 0 , \quad (T - 2I)(x_3) \neq 0$$

On the other hand since J is the equivalent matrix to matrix representation of T then

$$T(x_2) = x_1 + 2x_2 + 0x_3 + \cdots + 0x_8$$
$$T(x_3) = 0x_1 + x_2 + 2x_3 + 0x_4 + \cdots + 0x_8$$

then $(T - 2I)^2 (x_2) = (T - 2I)(T(x_2) - 2x_2) = (T - 2I)(x_1) = 0$ and $(T - 2I)^3$ $(x_3) = (T - 2I)^2 (T(x_3) - 2x_3) = (T - 2I)^2 (x_2) = 0$, therefore, although $(T - 2I)(x_2) \neq 0$ and $(T - 2I)(x_3) \neq 0$ it follows that

$$(T - 2I)^p (x_2) = (T - 2I)^p (x_3) = 0 \quad \text{if } p \geq 2$$

If p is the smallest positive integer such that $(T - \lambda I)^p (x) = 0$ then $(T - \lambda I)^{p-1}$ $(x) = y$ is an eigenvector corresponding to eigenvalue λ.

Definition 8.7 Let T a linear operator in a space vector V. A subspace W of V is called $T -$ cyclic subspace if there exists an element $v \in W$ such that W is equal to the generated subspace by $\{v, T(v), T^2(v), \ldots\}$. In this case is, it is said that W is generated by v.

Example 8.10 Let $T : \mathbb{R}^3 \to \mathbb{R}^3$ defined by $T(a, b, c) = (-b + c, a + c, 3c)$. The $T -$ cyclic subspace generated by $e_1 = (1, 0, 0)$ is determined as follows:

$$T(e_1) = T(1, 0, 0) = (0, 1, 0) = e_2$$
$$T^2(e_1) = T(T(e_1)) = T(e_2) = (-1, 0, 0) = -e_1$$

then

$$\mathscr{L}\left\{e_1, T(e_1), T^2(e_1), \ldots\right\} = \mathscr{L}\left\{e_1, e_2\right\}.$$

Definition 8.8 Let T and V as above definition. Let v a generalized eigenvector of T corresponding to eigenvalue λ. If p is the smallest positive integer such that $(T - \lambda I)^p(v) = 0$ then the set $\left\{(T - \lambda I)^{p-1}(v), (T - \lambda I)^{p-2}(v), \ldots, (T - \lambda I)(v), v\right\}$ is called a **cycle of generalized eigenvectors** of T corresponding to eigenvalue λ.

Theorem 8.7 *Let T a linear operator in V, γ a cycle of generalized eigenvectors of T corresponding to eigenvalue λ then:*

1. *The initial vector of γ is a eigenvector of T corresponding to eigenvalue λ and no other element of γ is a eigenvector of T.*
2. *γ is linearly independent.*
3. *Let β a base of V. Then β is a Jordan canonical base for V if and only if β is a disjoint union of cycles of generalized eigenvectors of T.*

Proof It is left to the reader as an exercise. $\square$

Definition 8.9 Let λ a eigenvalue of a linear operator T in V. The set given by:

$$K_\lambda = \left\{v \in V : (T - \lambda I)^p(v) = 0, \ p \in \mathbb{Z}^+\right\}$$

is called **generalized eigenspace** of T corresponding to λ such that $E_\lambda \subseteq K_\lambda$. If the algebraic multiplicity of λ is m then $\dim K_\lambda = m$, hence $\dim E_\lambda \leq \dim K_\lambda = m$.

8.2.2   Dot Diagram Method

As an aid to calculate J_i and the base β_i, we introduce an arrangement of points called **dot diagram**. Let suppose that β_i is a disjoint union of cycles $c_1, c_2, \ldots, c_{k_i}$ with lengths $p_1 \geq p_2 \geq \cdots \geq p_{k_i}$ respectively. The diagram is constructed as follows.

1. The arrangement consists in k_i columns. One column for each cycle.
2. From left to right, the column j has p_j points that correspond to elements of c_j in the following form. If x_j is the *vector terminal* of c_j the:

   - The top point corresponds to $(T - \lambda_i I)^{p_j - 1}(x_j)$.
   - The second point corresponds to $(T - \lambda_i I)^{p_j - 2}(x_j)$.
   - The last point (point of the bottom) of the column corresponds to x_j.

Finally, the associated diagram to β_i can be written as:

- $(T - \lambda_i I)^{p_1-1} (x_1)$ • $(T - \lambda_i I)^{p_2-1} (x_2)$ $\cdots$ • $(T - \lambda_i I)^{p_{k_i}-1} (x_{k_i})$
- $(T - \lambda_i I)^{p_1-2} (x_1)$ • $(T - \lambda_i I)^{p_2-2} (x_2)$ $\cdots$ • $(T - \lambda_i I)^{p_{k_i}-2} (x_{k_i})$

$$\vdots \qquad\qquad \vdots \qquad\qquad \vdots$$

$$\vdots \qquad\qquad \vdots \qquad\qquad \bullet \quad (T - \lambda_i I)(x_{k_i})$$

$$\bullet \quad (T - \lambda_i I)(x_2) \qquad \bullet \qquad x_{k_i}$$

- $(T - \lambda_i I)(x_1)$ • x_2
- x_1

where vector x_j always is a eigenvector. Note that dot diagram is for each λ_i.

Besides since $p_1 \geq p_2 \geq \cdots \geq p_{k_i}$, the columns of the dot diagram become shorter (or at least no longer) when we move from left to right. How many points are there in the first row of the diagram? The following theorem has the answer to this question.

Theorem 8.8 *Let r_j the number of points in the row j of a dot diagram for the base β_i, then*

$$r_1 = \dim V - \text{rank}\,(T - \lambda_i I)$$
$$r_j = \text{rank}\,(T - \lambda_i I)^{j-1} - \text{rank}\,(T - \lambda_i I)^j \ \text{ for } j > 1.$$

Proof It is left to the reader as an exercise. □

Example 8.11 In Example 8.4, we saw that matrix A given by

$$A = \begin{bmatrix} -3 & 3 & -2 \\ -7 & 6 & -3 \\ 1 & -1 & 2 \end{bmatrix}$$

is not diagonalizable. For this matrix, the eigenvalues are $\lambda_1 = 2$ with algebraic multiplicity 2 and $\dim E_{\lambda_1} = 1$, $\lambda_2 = 1$ with algebraic multiplicity 1 and $\dim E_{\lambda_2} = 1$. Besides $\dim E_{\lambda_1} < \dim K_{\lambda_1} = 2$. Let β_1 a base for K_{λ_1}, by Theorem 8.8,

$$r_1 = 3 - \text{rank}\,(A - 2I) = 3 - 2 = 1$$

Hence in the firs row there is one point. Since $\dim K_{\lambda_1} = 2$ there are only two linearly independent vectors, then in the second row there is one point too. Therefore the associated dot diagram with β_i is:

- $(A - 2I)(x_1)$
- x_1

where x_1 is a linearly independent eigenvector, then

$$J_1 = \begin{bmatrix} 2 & 1 \\ 0 & 2 \end{bmatrix}$$

After some calculations, the linearly independent eigenvector is obtained:

$$x_1 = \begin{bmatrix} 1 \\ 1 \\ -1 \end{bmatrix}$$

To find the generalized eigenvector v, the easiest method for practice cases is to calculate from the equation $(A - 2I)v = x_1$, i.e.,

$$\begin{bmatrix} -5 & 3 & -2 \\ -7 & 4 & -3 \\ 1 & -1 & 0 \end{bmatrix} \begin{bmatrix} v_1 \\ v_2 \\ v_3 \end{bmatrix} = \begin{bmatrix} 1 \\ 1 \\ -1 \end{bmatrix}$$

where it follows that:

$$v = \begin{bmatrix} 1 \\ 2 \\ 0 \end{bmatrix}$$

hence

$$\beta_1 = \left\{ \begin{bmatrix} 1 \\ 1 \\ -1 \end{bmatrix}, \begin{bmatrix} 1 \\ 2 \\ 0 \end{bmatrix} \right\}$$

then

$$K_{\lambda_1} = \mathscr{L} \left\{ \begin{bmatrix} 1 \\ 1 \\ -1 \end{bmatrix}, \begin{bmatrix} 1 \\ 2 \\ 0 \end{bmatrix} \right\}$$

On the other hand, $\dim E_{\lambda_2} = \dim K_{\lambda_2} = 1$. Let β_2 a base for K_{λ_2}, since $\dim K_{\lambda_2} = 1$ there is only one linearly independent vector and the dot diagram is:

$$\bullet \, x_1$$

then $J_2 = (\lambda_2) = (1)$. Solving $(A - I)v = 0$ it is obtained

$$v = \begin{bmatrix} 1 \\ 2 \\ 1 \end{bmatrix}$$

then

$$\beta_2 = \left\{ \begin{bmatrix} 1 \\ 2 \\ 1 \end{bmatrix} \right\}$$

then

$$K_{\lambda_2} = \mathscr{L}\left\{\begin{bmatrix} 1 \\ 2 \\ 1 \end{bmatrix}\right\}$$

If β is the Jordan canonical base, by Theorem 8.7,

$$\beta = \beta_1 \cup \beta_2 = \left\{\begin{bmatrix} 1 \\ 1 \\ -1 \end{bmatrix}, \begin{bmatrix} 1 \\ 2 \\ 0 \end{bmatrix}, \begin{bmatrix} 1 \\ 2 \\ 1 \end{bmatrix}\right\}$$

and the Jordan canonical form is

$$J = \begin{bmatrix} 2 & 1 & \vdots & 0 \\ 0 & 2 & \vdots & 0 \\ \cdots & \cdots & \cdots & \cdots \\ 0 & 0 & \vdots & 1 \end{bmatrix}$$

Then the transformation matrix Q is:

$$Q = \begin{bmatrix} 1 & 1 & 1 \\ 1 & 2 & 2 \\ -1 & 0 & 1 \end{bmatrix}$$

Hence $J = Q^{-1}AQ$.

Exercise 8.1 If the characteristic polynomial of a matrix B is $(\lambda + 3)^2 (\lambda - 2)^2 (\lambda - 4)$, find all possible Jordan canonical forms.

References

1. Hoffman, K., Kunze, R.: Linear Algebra. Prentice-Hall (1971)
2. Axler, S.: Linear Algebra Done-Right. Springer-Verlag (1997)
3. Lipschutz, S.: Linear Algebra. McGraw Hill (2009)
4. Anton, H.: Elementary Linear Algebra. Wiley (2010)
5. Grossman, S.: Elementary Linear Algebra. Wadsworth Publishing (1987)
6. Whitelaw, T.: Introduction to Linear Algebra. Chapman & Hall/CRC (1992)
7. kolman, B.: Introductory Linear Algebra an Applied First Course. Pearson Education (2008)
8. Halmos, P.R.: Finite-Dimensional Vector Spaces. Courier Dover Publications (2017)
9. Nef, W.: Linear Algebra. Dover Publications (1967)
10. Lang, S.: Linear Algebra. Springer (2004)
11. Nering, E.: Linear Algebra and Matrix Theory. Wiley (1970)

Chapter 9
Differential Equations

Abstract The purpose of this chapter is to present several methods to solve differential equations, this chapter begins with a motivation based on physical phenomena that can be represented with differential equations, after that we introduce definitions and methods for first order and second order linear differential equations, various exercises are given to the reader to better illustrate the contents of this chapter.

9.1 Motivation: Some Physical Origins of Differential Equations

9.1.1 Free Fall

Free fall is any motion of a body where gravity is the only force acting upon it. Consider the free fall of a body with mass m and $y(0) = 0$, $y'(0) = 0$ (Fig. 9.1). By second law of Newton we have:

$$+ \downarrow \sum F_y = ma$$
$$mg = ma$$
$$g = a = \frac{d^2y}{dt^2}$$

since $\dfrac{d^2y}{dt^2} = \dfrac{dv}{dt}$ then $\displaystyle\int dv = \int g dt$ hence:

$$v = gt + k_1$$

Integrating the above equation, and taking into account that $v = \dfrac{dy}{dt}$:

$$y = \int v dt = \int (gt + k_1)dt = \frac{1}{2}gt^2 + k_1t + k_2$$

© Springer Nature Switzerland AG 2019

R. Martínez-Guerra et al., *Algebraic and Differential Methods for Nonlinear Control Theory*, Mathematical and Analytical Techniques with Applications to Engineering, https://doi.org/10.1007/978-3-030-12025-2_9

Fig. 9.1 Free fall

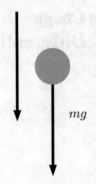

mg

Finally, by initial conditions of the problem:

$$k_1 = y'(0) - g(0) = 0$$
$$k_2 = y(0) - \frac{1}{2}g(0)^2 + k_1(0) = 0$$

therefore

$$y = \frac{1}{2}gt^2$$

9.1.2 Simple Pendulum Problem

Now, according to Fig. 9.2, it is not difficult to obtain the following analysis based in kinetic and potential energy for the simple pendulum that consists in a mass m attached to a rope with length L. From points 1 and 2 in the figure:

$$\frac{1}{2}mv^2 + mgL\,(1 - \cos\theta) = mgL\,(1 - \cos\alpha)$$
$$\frac{1}{2}v^2 + gL\,(1 - \cos\theta) = gL\,(1 - \cos\alpha)$$
$$v^2 + 2gL\,(1 - \cos\theta) = 2gL\,(1 - \cos\alpha)$$

since $s = \theta L$ then $v = \dfrac{ds}{dt} = L\dfrac{d\theta}{dt}$, we have:

$$L\left(\frac{d\theta}{dt}\right)^2 + 2g\,(\cos\alpha - \cos\theta) = 0$$

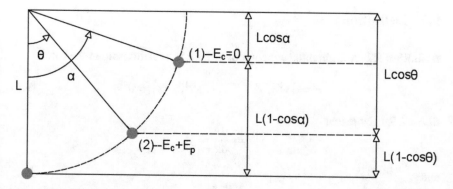

Fig. 9.2 Simple pendulum

9.1.3 Friction Problem

According to Fig. 9.3, we have:

$$mg - F_f = m\frac{d^2y}{dt^2}$$

then

$$mg - \mu Kv = m\frac{d^2y}{dt^2}$$

$$mg - \bar{K}\frac{dy}{dt} = m\frac{d^2y}{dt^2}$$

where $\bar{K} = \mu K$. Finally we have:

$$\frac{d^2y}{dt^2} + \frac{\bar{K}}{m}\frac{dy}{dt} - g = 0$$

Fig. 9.3 Friction problem

9.2   Definitions

Definition 9.1 A differential equation of order n is a relationship of the form:

$$F = \left(y^{(n)}, y^{(n-1)}, \ldots, y^{(1)}, y, x\right) = 0 \tag{9.1}$$

Remark 9.1 Note that

$$y^{(1)} = \frac{dy}{dx} = y'$$

and

$$y^{(2)} = \frac{d^2 y}{dx^2} = y''$$

$y' = \dot{y} = y^{(1)}, y'' = \ddot{y} = y^{(2)}$ are commonly used to represent the first and second derivatives with respect to the time respectively.

Definition 9.2 It is said to be $f : (a, b) \to \mathbb{R}$, $f \in \mathscr{C}^n$, is a solution of (9.1), if for all $x \in (a, b)$:

$$F = \left(f^{(n)}(x), f^{(n-1)}(x), \ldots, f'(x), f(x), x\right) = 0 \tag{9.2}$$

Example 9.1 1. The relation $F = y'' - y = 0$ is a differential equation of second order. Note that

$$f_1(x) = e^x$$
$$f_2(x) = k_1 e^x, k_1 \in \mathbb{R}$$
$$f_3(x) = e^x + e^{-x}$$
$$f_4(x) = k_2 e^x + k_3 e^{-x}, k_2, k_3 \in \mathbb{R}$$

are solutions of F.

2. The solutions of the differential equation of second order $y'' - 2y' + y = 0$ are given by $y = f(x) = xe^x$ and $y = e^x$.
3. $y' = \cos x$ is a nonlinear differential equation of first order.
4. $y'' + k^2 y = 0$ is a differential equation of second order.
5. $\left(x^2 + y^2\right) dx + 2xy dy = 0$ is a nonlinear differential equation of first order.
6. $\left[\frac{d^3 w}{dx^2}\right]^3 - xy\frac{dw}{dx} + w = 0$ is a differential equation of second order and third degree.

9.3 Separable Differential Equations

Definition 9.3 The equation of first order $F(y', y, x) = 0$ is called **normal**, if the first derivative can be obtained, i.e. [1–3]:

$$\frac{dy}{dx} = y' = f(x, y) \tag{9.3}$$

Definition 9.4 The Eq. (9.3) is said to be **separable differential equation** if we can write it in the following form:

$$N(y)dy = M(x)dx \tag{9.4}$$

Solving separable differential equations is easy. We first rewrite the differential equation as the follows

$$N(y)dy = M(x)dx$$

Then, integrating in both sides, we have:

$$\int N(y)dy = \int M(x)dx$$

Example 9.2 1. $x\,dx - y\,dy = 0$. In this case, the differential equation can be write in the following form $x\,dx = y\,dy$, then integrating in both sides we have that $x^2 + c = y^2$, where it follows that $y = \pm\sqrt{x^2 + c}$.

2. $x\left(1 + y^2\right)dx - y\left(1 + x^2\right)dy = 0$. For this DE we have:

$$x\left(1 + y^2\right)dx - y\left(1 + x^2\right)dy = 0$$
$$y\left(1 + x^2\right)dy = x\left(1 + y^2\right)dx$$
$$\int \frac{x\,dx}{1 + x^2} = \int \frac{y\,dy}{1 + y^2}$$
$$\ln(1 + y^2) = \ln(1 + x^2) + C$$
$$y^2 = C_1(1 + x^2) - 1$$
$$y = \pm\sqrt{C_1(1 + x^2) - 1}$$

3. $\frac{dy}{dx} + y^2 e^x = y^2$. This differential equation is separable, then:

$$\frac{dy}{dx} + y^2 e^x = y^2$$

$$\int \frac{dy}{y^2} = \int (1 - e^x) dx$$

$$-y^{-1} = x - e^x + C$$

$$y = \frac{1}{e^x - x + C}$$

4. $\dfrac{dy}{dx} = \cos^2(y)\sin(x)$.

5. $\dfrac{dy}{dx} = \dfrac{1+y}{1-x}$

6. $ay\ln(x) = \frac{dy}{dx} - ay, a \in \mathbb{R}$

The solution for differential equations 4, 5 and 6 are left to the reader as an exercise.

9.4   Homogeneous Equations

Definition 9.5 A function $f(x, y)$ is said to be **homogeneous** of degree n ($n = 0, 1, \ldots$) if for all $t > 0$ [4, 5]:

$$f(tx, ty) = t^n f(x, y)$$

Example 9.3   1. If $f(x, y) = x^3 + xy^2$ then $f(tx, ty) = (tx)^3 + (tx)(ty)^2 = t^3 (x^3 + xy^2) = t^3 f(x, y)$. Hence $f(t, x)$ is homogeneous of degree 3.

2. $f(x, y) = 1 + \sin\left(\dfrac{x}{y}\right)$ is homogeneous of degree 0:

$$f(tx, ty) = 1 + \sin\left(\frac{tx}{ty}\right) = 1 + \sin\left(\frac{x}{y}\right) = t^0 f(x, y)$$

3. In the same manner as above, it is not difficult to prove that $f(x, y) = x + \sqrt{x^2 + y^2}$ is homogeneous of degree 1. The proof is left to the reader.

Definition 9.6 An equation $\frac{dy}{dx} = f(x, y)$ is said to be **homogeneous** if f is homogeneous of zero degree.

Example 9.4

$$\frac{dy}{dx} = \frac{x + y}{3x - 2y}$$

is homogeneous because $f(x, y) = \frac{x+y}{3x-2y}$ is a homogenous function of degree zero:

$$f(tx, ty) = \frac{tx + ty}{3tx - 2ty} = \frac{t(x + y)}{t(3x - 2y)} = t^0 f(x, y)$$

Proposition 9.1 *Let $M(x, y)$ and $N(x, y)$ homogeneous functions of the same degree (say n) then the differential equation*

$$M(x, y)dx + N(x, y)dy = 0$$

is homogeneous.

Proof The proof is immediate and it is left to the reader. □

Now, let consider $f(x, y)$ a homogeneous equation of zero degree, i.e.,

$$f(tx, ty) = t^0 f(x, y) = f(x, y)$$

if

$$f\left(1, \frac{y}{x}\right) = f(x, y), \quad t = \frac{1}{x}, x \neq 0$$

then for a differential equation $\frac{dy}{dx} = f(1, \frac{y}{x})$, let define $z = \frac{y}{x}$, then

$$y = xz \tag{9.5}$$

Deriving (9.5), we have

$$\frac{dy}{dx} = \frac{d}{dx}(xz) = x\frac{dz}{dx} + z\frac{dx}{dx} = x\frac{dz}{dx} + z = f(1, z)$$

then

$$\frac{dz}{f(1, z) - z} = \frac{dx}{x}$$

The problem is solved as a separable differential equation problem.

Example 9.5 Given

$$y' = \frac{2x + y}{x}$$

We define $z = \frac{y}{x}$. Then

$$x\frac{dz}{dx} + z = 2 + z$$

$$x\frac{dz}{dx} = 2$$

$$\int dz = 2\int \frac{dx}{x}$$

$$z = 2\ln x + k, k \in \mathbb{R}$$

Returning to the original variable $\frac{y}{x} = 2 \ln x + k$. Finally:

$$y = \ln x^{2x} + kx$$

Exercise 9.1 Solve the following differential equations.

1. $\dfrac{dy}{dx} = \dfrac{x+y}{x-y}$
2. $\sqrt{x^2 + y^2}\, dx = y\, dy$
3. $\left(x^2 + y^2\right) dx\; xy\, dy = 0$
4. $\dfrac{dy}{dx} = \dfrac{1+y}{1-x}$
5. $ay \ln(x) = \frac{dy}{dx} - ay, a \in \mathbb{R}$

9.5   Exact Equations

Let $F(x, y)$, then $dF = \dfrac{\partial F}{\partial x} dx + \dfrac{\partial F}{\partial y} dy$. If F is a constant function, then

$$M(x, y)dx + N(x, y)dy = \frac{\partial F}{\partial x} dx + \frac{\partial F}{\partial y} dy = 0 \tag{9.6}$$

hence $\dfrac{\partial F}{\partial x} = M(x, y), \dfrac{\partial F}{\partial y} = N(x, y)$. The differential equation (9.6) is said to be **exact** if there exists $F(x, y)$ such that:

$$M(x, y) = \frac{\partial F(x, y)}{\partial x} \quad \text{and} \quad N(x, y) = \frac{\partial F(x, y)}{\partial y}$$

In this case, the Eq. (9.6) can be written as:

$$dF = M(x, y)dx + N(x, y)dy = 0$$

and its solutions are given by $F(x, y) = $ constant.

Example 9.6 The differential equation

$$ye^x dx + \left(e^x + 2y\right) dy = 0$$

is exact. Let $F(x, y) = ye^x + y^2 = c$ with c a constant, then

$$\frac{\partial F(x, y)}{\partial x} = ye^x \quad \text{and} \quad \frac{\partial F(x, y)}{\partial y} = e^x + 2y$$

Now let us check the following problem. If $M(x, y)dx + N(x, y)dy$ is exact, then

$$\frac{\partial M}{\partial y} = \frac{\partial N}{\partial x}$$

if and only if dx and dy are continuous. Reciprocally if $M(x, y)dx + N(x, y)dy$ is a differential equation such that:

$$\frac{\partial M}{\partial y} = \frac{\partial N}{\partial x}$$

then the equation is exact. To solve this problem, let us suppose the equation $\frac{\partial M}{\partial y} = \frac{\partial N}{\partial x}$, then there exists $F(x, y)$ such that

$$M = \frac{\partial F}{\partial x}$$

therefore

$$F(x, y) = \int M(x, y)\, dx + g(y) \tag{9.7}$$

but $\frac{\partial F}{\partial y} = N(x, y)$, i.e.

$$\frac{\partial}{\partial y}\left[\int M(x, y)\, dx\right] + g'(y) = N(x, y)$$

then

$$g'(y) = N(x, y) - \frac{\partial}{\partial y}\int M(x, y)\, dx$$

Finally

$$g(y) = \int\left[N(x, y) - \frac{\partial}{\partial y}\int M(x, y)\, dx\right]dy$$

Now, the substitution in (9.7) of $g(y)$ is useful to find the function $F(x, y)$.

Example 9.7 In the differential equation

$$ye^x dx + \left(e^x + 2y\right) dy = 0$$

let $M = ye^x$ and $N = e^x + 2y$, so that $\frac{\partial M}{\partial y} = e^x$ and $\frac{\partial N}{\partial x} = e^x$. Therefore the given differential equation is exact. From (9.7):

$$F(x, y) = \int ye^x\, dx + g(y) = ye^x + g(y)$$

In addition

$$N = \frac{\partial F(x, y)}{\partial y} = e^x + g'(y) = e^x + 2y$$

so that $g'(y) = 2y$ and $g(y) = y^2 + k$. Finally

$$F(x, y) = ye^x + y^2 = \text{constant}$$

Exercise 9.2 Solve the following differential equations.

1. $\left(x + \dfrac{2}{y}\right) dy + y dx = 0$
2. $\left(y - x^3\right) dx + \left(x + y^3\right) dy = 0$
3. $(y + y \cos(xy)) dx + (x + x \cos(xy)) dy = 0$
4. $(\sin x \sin y - xe^y) dy = (e^y + \cos x \cos y) dx$
5. $e^y dx + (xe^y + 2y) dy = 0$

9.6   Linear Differential Equations

Definition 9.7 A differential equation of order n is said to be **linear** if has the following form:

$$a_n(x) y^{(n)} + a_{n-1}(x) y^{(n-1)} + \cdots + a_1(x) y^{(1)} + a_0(x) y = f(x) \qquad (9.8)$$

As well as, the differential equation (9.8) is said to be **homogeneous** [6, 7] if $f(x) = 0 \, \forall \, x$ and the differential equation (9.8) is said to be **with constants coefficients** if $a_n(x), a_{n-1}(x), \ldots, a_1(x), a_0(x)$ are constants functions.

Example 9.8   1. $2xy'' - xy' = 2$ is a differential equation of second order, linear, with variables coefficients and non homogeneous.
2. $y^{(3)} - y^{(2)} + xy = 0$ is a differential equation of third order, linear, with variables coefficients and homogeneous.
3. $y^{(2)} + \lambda^2 y = \cos(\omega x)$ is a linear differential equation of second order with constants coefficients and non homogeneous.

Now we consider three cases. The **first case** is given by the following differential equation

$$ay' + by = 0 \quad a, b \text{ are constants, } a \neq 0 \qquad (9.9)$$

Solving (9.9) we have:

$$a\frac{dy}{dx} = -by$$

$$\frac{dy}{y} = -\frac{b}{a}dx$$

$$\int \frac{dy}{y} = -\frac{b}{a}\int dx$$

$$\ln y = -\frac{b}{a}x + C_1$$

$$y = C_2 e^{-\frac{b}{a}x}$$

with $C_1, C_2 = e^{C_1}$ constants.

The **second case** is given by the following differential equation

$$ay' + by = f(x) \quad a, b \text{ are constants}, a \neq 0 \tag{9.10}$$

Rewriting (9.10) we have:

$$y' + \lambda y = h(x) \tag{9.11}$$

where $\lambda = \frac{b}{a}, h(x) = \frac{1}{a}f(x), a \neq 0$. On the other hand, we have

$$\left(e^{\lambda x}y\right)' = e^{\lambda x}y' + e^{\lambda x}\lambda y = e^{\lambda x}\left(y' + \lambda y\right)$$

From (9.11) multiplying by $e^{\lambda x}$:

$$e^{\lambda x}\left(y' + \lambda y\right) = e^{\lambda x}h(x)$$

so that we have

$$\left(e^{\lambda x}y\right)' = e^{\lambda x}h(x)$$

Finally,

$$e^{\lambda x}y = \int e^{\lambda x}h(x)\,dx + C$$

and the solution of (9.10) is given by:

$$y = e^{-\lambda x}\int e^{\lambda x}h(x)\,dx + e^{-\lambda x}C \tag{9.12}$$

Example 9.9 Given the differential equation

$$y' + 2y = e^{5x}$$

Applying the methodology previously explained, we have

$$e^{2x}\left(y' + 2y\right) = e^{7x}$$
$$\left(e^{2x}y\right)' = e^{7x}$$
$$e^{2x}y = \frac{1}{7}e^{7x} + K$$
$$y(x) = \frac{1}{7}e^{5x} + Ke^{-2x}$$

Note that $a = 1, b = 2, f(x) = e^{5x}$, so that $\lambda = \frac{b}{a} = 2$ and $h(x) = \frac{1}{a}f(x) = e^{5x}$ and from (9.12), the solution of the differential equation is:

$$y(x) = \frac{1}{7}e^{5x} + Ke^{-2x}$$

and K is the integration constant.

Finally the **third case** is given by the following differential equation:

$$y' + b(x)y = f(x) \tag{9.13}$$

Notice that

$$\left(e^{B(x)}y\right)' = e^{B(x)}\left(y' + B'(x)y\right)$$

so that it is proposed $B'(x) = b(x)$. From (9.13) multiplying by $e^{B(x)}$, we obtain:

$$e^{B(x)}\left(y' + b(x)y\right) = \left(e^{B(x)}y\right)' = e^{B(x)}f(x)$$

therefore

$$e^{B(x)}y = \int e^{B(x)}f(x)\,dx + C$$

and the solution of (9.13) is:

$$y = e^{-B(x)}\int e^{B(x)}f(x)\,dx + Ce^{-B(x)}, \text{ with } B(x) = \int b(x) \tag{9.14}$$

Example 9.10 To solve the following differential equation

$$y' + y\sin x = \sin x$$

note that $B'(x) = b(x) = \sin x$, then $B(x) = \int \sin x = -\cos x + k_1$ and $f(x) = \sin x$. Then from (9.14), we have:

$$y = e^{\cos x - k_1} \int e^{-\cos x + k_1} \sin x \, dx + C e^{\cos x - k_1}$$

$$= e^{-k_1} e^{\cos x} \int e^{k_1} e^{-\cos x} \sin x \, dx + C e^{\cos x - k_1}$$

$$= e^{\cos x} \int e^{-\cos x} \sin x \, dx + C e^{\cos x - k_1}$$

$$= e^{\cos x} \left[e^{-\cos x} + k_2 \right] + C e^{\cos x - k_1}$$

$$= 1 + k_2 e^{\cos x} + C e^{\cos x - k_1}$$

$$y = 1 + k_3 e^{\cos x}$$

where k_1, k_2 are integration constants and $k_3 = k_2 + \beta$, $\beta = C e^{-k_1}$.

9.7 Homogeneous Second Order Linear Differential Equations

In the homogeneous second order linear differential equation with constants coefficients a_0, a_1:

$$y'' + a_1 y' + a_0 y = 0 \tag{9.15}$$

suppose that $y = e^{\lambda x}$ is a solution. The derivatives of the proposed solution are given by

$$y' = \lambda e^{\lambda x}, \quad y'' = \lambda^2 e^{\lambda x}$$

Substituting in (9.15), we have:

$$\lambda^2 e^{\lambda x} + a_1 \lambda e^{\lambda x} + a_0 e^{\lambda x} = e^{\lambda x} \left[\lambda^2 + a_1 \lambda + a_0 \right] = 0$$

If λ is a root of the equation $\lambda^2 + a_1 \lambda + a_0 = 0$ then the function $e^{\lambda x}$ is a solution of (9.15). The polynomial $p(\lambda) = \lambda^2 + a_1 \lambda + a_0$ is called the **characteristic polynomial** and the equation $\lambda^2 + a_1 \lambda + a_0 = 0$ is the **characteristic equation** of (9.15) [2, 5, 7]. The roots of the characteristic equation are called **characteristic roots**.

Theorem 9.1 *Given the differential equation*

$$y'' + a_1 y' + a_0 y = 0 \tag{9.16}$$

We consider the roots of characteristic equation:

$$\lambda^2 + a_1 \lambda + a_0 = 0 \tag{9.17}$$

1. If $r_1 \neq r_2$, let define $\varphi_1(x) = e^{r_1 x}$, $\varphi_2(x) = e^{r_2 x}$.
2. If $r_1 = r_2 = r$, let define $\varphi_1(x) = e^{r x}$, $\varphi_2(x) = x e^{r x}$.

then in any case, the functions $\varphi_1(x)$ and $\varphi_2(x)$ are solutions. In addition, any solution $\varphi(x)$ of differential equation is of the form:

$$y(x) = c_1\varphi_1(x) + c_2\varphi_2(x)$$

where c_1, c_2 are constants.

Proof We suppose that the roots of equation $\lambda^2 + a_1\lambda + a_0 = 0$ are equals ($r_1 = r_2 = r$). $\varphi_2(x) = xe^{rx}$ is a solution of (9.16), i.e., $\varphi_2'(x) = (rx + 1)e^{rx}$, $\varphi_2''(x) = r(rx + 2)e^{rx}$. Substituting in (9.16) we have $r(rx + 2)e^{rx} + a_1(rx + 1)e^{rx} + a_0xe^{rx}$ then $e^{rx}\left[x(r^2 + a_1r + a_0) + 2r + a_1\right]$. Since r is a root, this yields to

$$e^{rx}\left[2r + a_1\right] \tag{9.18}$$

Since $r = r_1 = r_2, r = -\dfrac{1}{2}a_1$. Substituting in (9.18), we have

$$e^{rx}\left[2\left(-\frac{1}{2}a_1\right) + a_1\right] = 0$$

Hence φ_2 is a solution of (9.16). If φ_1 and φ_2 are solutions of (9.16) then $\varphi = c_1\varphi_1 + c_2\varphi_2$ is a solution of (9.16), since

$$\left(c_1\varphi_1'' + c_2\varphi_2''\right) + a_1\left(c_1\varphi_1' + c_2\varphi_2'\right) + a_0\left(c_1\varphi_1 + c_2\varphi_2\right)$$
$$= c_1\left(\varphi_1'' + a_1\varphi_1' + a_0\varphi_1\right) + c_2\left(\varphi_2'' + a_1\varphi_2' + a_0\varphi_2\right) = 0$$

Thus the linear combination $\varphi = c_1\varphi_1 + c_2\varphi_2$ is a solution of (9.16).   □

Example 9.11   1.   Let
$$y'' - 4y' + 4y = 0 \tag{9.19}$$

Its characteristic equation is given by

$$\lambda^2 - 4\lambda + 4 = (\lambda - 2)^2 = 0$$

and its roots are given by $\lambda = 2$ with multiplicity $m = 2$. Therefore the solutions (according to **Theorem** 9.1) are given by

$$\varphi_1(x) = e^{2x}, \varphi_2(x) = xe^{2x}$$

and the complete solution for (9.19) is

$$y(x) = c_1\varphi_1(x) + c_2\varphi_2(x) = c_1e^{2x} + c_2xe^{2x}$$

2. The differential equation given by $y'' - 3y' + 2y = 0$ has a characteristic equation $\lambda^2 - 3\lambda + 2 = (\lambda - 1)(\lambda - 2) = 0$ with roots $\lambda_1 = 1$ and $\lambda_2 = 2$. Thus the functions $y_1 = e^x$ and $y_2 = e^{2x}$ are solutions of the linear homogeneous equation.

Exercise 9.3 Solve the following differential equations.

1. $y'' - 4y' + 5y = 0$
2. $y'' + 8y = c, c \in \mathbb{R}$
3. $2y'' + 2y' + 3y = 0$
4. The derivative of any solution of $y'' + a_1 y' + a_0 y = 0$ is a solution of the equation as well.

9.8 Variation of Parameters Method

Given the non-homogeneous differential equation of second order with constants coefficients:

$$y'' + a_1 y' + a_0 y = b(x) \tag{9.20}$$

if $b(x) = 0$, then (9.20) is a homogeneous equation. The solution of (9.20) is given in the following result.

Theorem 9.2 *If $\varphi_1(x)$ and $\varphi_2(x)$ are solutions of the homogeneous equation associate to (9.20) and $\varphi_p(x)$ is any solution of (9.20), i.e., a particular solution, then the general solution of (9.20) is:*

$$y(x) = c_1 \varphi_1(x) + c_2 \varphi_2(x) + \varphi_p(x)$$

where c_1, c_2 are constants.

The methodology to solve (9.20) is described as follows. The constants c_1 and c_2 are replaced by unknown functions $\mu_1(x)$ and $\mu_2(x)$ such that

$$y = \mu_1 \varphi_1 + \mu_2 \varphi_2 \tag{9.21}$$

is a solution of (9.20). Therefore, the main objective is to find $\mu_1(x)$ and $\mu_2(x)$ such that (9.21) is a solution of the non-homogeneous equation (9.20).

From (9.21), we have

$$y' = \mu_1 \varphi_1' + \mu_1' \varphi_1 + \mu_2 \varphi_2' + \mu_2' \varphi_2$$
$$y'' = 2\mu_1' \varphi_1' + 2\mu_2' \varphi_2' + \mu_1 \varphi_1'' + \mu_1'' \varphi_1 + \mu_2 \varphi_2'' + \mu_2'' \varphi_2$$

Substituting the above equations in (9.20), we have

$$\mu_1 \varphi_1'' + a_1 \mu_1 \varphi_1' + a_0 \mu_1 \varphi_1 = 0$$
$$\mu_2 \varphi_2'' + a_1 \mu_2 \varphi_2' + a_0 \mu_2 \varphi_2 = 0$$

since φ_1 and φ_2 are solutions of the homogeneous equation, then:

$$2\mu_1'\varphi_1' + 2\mu_2'\varphi_2' + \mu_1''\varphi_1 + \varphi_2\mu_2'' + a_1\mu_1'\varphi_1 + a_1\mu_2'\varphi_2 = b(x) \tag{9.22}$$

Furthermore, if $\varphi_1\mu_1' + \varphi_2\mu_2' = 0$, then its derivative is $\varphi_1\mu_1'' + \varphi_1'\mu_1' + \varphi_2\mu_2'' + \varphi_2'\mu_2' = 0$, finally from (9.22) we have

$$\mu_1'\varphi_1' + \mu_2'\varphi_2' = b(x)$$

Note that we have to solve a system of equations given by:

$$\left. \begin{array}{l} \varphi_1\mu_1' + \varphi_2\mu_2' = 0 \\ \varphi_1'\mu_1' + \varphi_2'\mu_2' = b(x) \end{array} \right\} \tag{9.23}$$

where μ_1', μ_2' are the variables to determine.

Definition 9.8 Let the system (9.23), the **Wronskian** (Wronsky determinant) is defined by:

$$W(\varphi_1, \varphi_2) = \begin{vmatrix} \varphi_1 & \varphi_2 \\ \varphi_1' & \varphi_2' \end{vmatrix} = \varphi_1\varphi_2' - \varphi_2\varphi_1' \neq 0 \tag{9.24}$$

Finally, the solution of (9.23) is:

$$\mu_1' = \frac{\begin{vmatrix} 0 & \varphi_2 \\ b(x) & \varphi_2' \end{vmatrix}}{W(\varphi_1, \varphi_2)} = \frac{-b(x)\varphi_2}{W(\varphi_1, \varphi_2)} \tag{9.25}$$

$$\mu_2' = \frac{\begin{vmatrix} \varphi_1 & 0 \\ \varphi_1' & b(x) \end{vmatrix}}{W(\varphi_1, \varphi_2)} = \frac{b(x)\varphi_1}{W(\varphi_1, \varphi_2)} \tag{9.26}$$

and the functions μ_1, μ_2 are given by the integration from (9.25) and (9.26)

$$\mu_1 = -\int \frac{b(x)\varphi_2}{W(\varphi_1, \varphi_2)} \quad , \quad \mu_2 = \int \frac{b(x)\varphi_1}{W(\varphi_1, \varphi_2)} \tag{9.27}$$

$\square$

Example 9.12 Let
$$y'' - 2y' + y = 2x$$

Note that the characteristic equation associate to the homogeneous differential equation is $\lambda^2 - 2\lambda + 1 = 0$. The root of this equation is $\lambda = 1$ with multiplicity two, then

$$\begin{array}{ll} \varphi_1(x) = e^x & \varphi_2(x) = xe^x \\ \varphi_1'(x) = e^x & \varphi_2'(x) = xe^x + e^x \end{array}$$

and the Wronskian for this problem is given by

$$W(\varphi_1, \varphi_2) = \begin{vmatrix} e^x & xe^x \\ e^x & xe^x + e^x \end{vmatrix} = e^x(xe^x + e^x) - xe^x e^x = xe^{2x} + e^{2x} - xe^{2x} = e^{2x} \neq 0$$

According to the methodology described above, we have

$$\mu_1 = -\int \frac{2x\,xe^x}{e^{2x}}\,dx = -2\int x^2 e^{-x}\,dx = 2e^{-x}(x^2 + 2x + 2)$$

$$\mu_2 = \int \frac{2x\,e^x}{e^{2x}}\,dx = 2\int xe^{-x}\,dx = -2e^{-x}(x+1)$$

Finally

$$\varphi(x) = \mu_1 \varphi_1 + \mu_2 \varphi_2 = 2x + 4$$

Exercise 9.4 Solve the following differential equations.

1. $y'' + 4y = \tan 2x$
2. $y'' + 2y' + y = e^{-x}\ln x$
3. $y'' - 2y' - 3y = xe^{-x}$
4. $y'' + 2y' + 5y = e^{-x}\sec 2x$

9.9 Initial Value Problem

Let $x_0 \in \mathbb{R}$, $\alpha, \beta \in \mathbb{R}$. To find a solution φ of the Eq. (9.15) such that

$$\varphi(x_0) = \alpha \quad , \quad \varphi'(x_0) = \beta$$

any solution that satisfies the conditions, can be written as

$$y = c_1 \varphi_1 + c_2 \varphi_2$$

Theorem 9.3 *Let $x_0, \alpha, \beta \in \mathbb{R}$, there exist unique constants c_1, c_2 such that the function*

$$\varphi(x) = c_1 \varphi_1(x) + c_2 \varphi_2(x), \quad \varphi(x_0) = \alpha, \varphi'(x_0) = \beta$$

where φ_1, φ_2 are given by

1. *$\varphi_1(x) = e^{r_1 x}$, $\varphi_2(x) = e^{r_2 x}$ if $r_1 \neq r_2$.*
2. *$\varphi_1(x) = e^{rx}$, $\varphi_2(x) = xe^{rx}$ if $r_1 = r_2 = r$.*

Proof Let c_1 and c_2 such that the following system

$$c_1 \varphi_1(x_0) + c_2 \varphi_2(x_0) = \alpha = \varphi(x_0) \Big\}$$
$$c_1 \varphi_1'(x_0) + c_2 \varphi_2'(x_0) = \beta = \varphi'(x_0) \Big\}$$

(9.28)

has a unique solution for cases 1 and 2. This condition is satisfied by

$$W\left(\varphi_1, \varphi_2\right) = \begin{vmatrix} \varphi_1(x_0) & \varphi_2(x_0) \\ \varphi_1'(x_0) & \varphi_2'(x_0) \end{vmatrix} = \varphi_1(x_0) \varphi_2'(x_0) - \varphi_2(x_0) \varphi_1'(x_0) \neq 0$$

For the case 1, the Wroskian is given by

$$W\left(\varphi_1, \varphi_2\right) = \begin{vmatrix} e^{r_1 x} & e^{r_2 x} \\ r_1 e^{r_1 x} & r_2 e^{r_2 x} \end{vmatrix} = e^{r_1 x} r_2 e^{r_2 x} - e^{r_2 x} r_1 e^{r_1 x} = (r_2 - r_1) e^{(r_1 + r_2)x} \neq 0$$

and for the case 2, the Wroskian is given by

$$W\left(\varphi_1, \varphi_2\right) = \begin{vmatrix} e^{rx} & x e^{rx} \\ r e^{rx} & (1 + rx) e^{rx} \end{vmatrix} = e^{2rx} \neq 0$$

therefore the solution is unique and there exist constants c_1, c_2 such that

$$\varphi = c_1 \varphi_1(x) + c_2 \varphi_2(x) \qquad\qquad \square$$

Proposition 9.2 Abel's Differential Equation Identity. *If $\varphi_1(x)$ and $\varphi_2(x)$ are solutions of:*

$$y'' + a_1 y' + a_0 y = 0 \tag{9.29}$$

then there exist A such that

$$W(\varphi_1, \varphi_2) = A e^{-a_1 x} \tag{9.30}$$

Proof By hypothesis $\varphi_1(x)$ and $\varphi_2(x)$ are solutions of (9.29), i.e.

$$\varphi_1'' + a_1 \varphi_1' + a_0 \varphi_1 = 0 \quad , \quad \varphi_2'' + a_1 \varphi_2' + a_0 \varphi_2 = 0$$

so that

$$\varphi_1 \left(\varphi_2'' + a_1 \varphi_2' + a_0 \varphi_2 \right) - \varphi_2 \left(\varphi_1'' + a_1 \varphi_1' + a_0 \varphi_1 \right) = 0$$

then

$$\left(\varphi_1 \varphi_2'' - \varphi_2 \varphi_1'' \right) + a_1 \left(\varphi_1 \varphi_2' - \varphi_1' \varphi_2 \right) = 0 \tag{9.31}$$

On the other hand, we have

$$W\left(\varphi_1, \varphi_2\right) = \begin{vmatrix} \varphi_1 & \varphi_2 \\ \varphi_1' & \varphi_2' \end{vmatrix} = \varphi_1 \varphi_2' - \varphi_2 \varphi_1'$$

and its derivative is given by

$$\frac{d}{dx} W (\varphi_1, \varphi_2) = \varphi_1 \varphi_2'' - \varphi_2 \varphi_1''$$

Note that (9.31) can be written as

$$\frac{d}{dx} W (\varphi_1, \varphi_2) + a_1 W (\varphi_1, \varphi_2) = 0 \tag{9.32}$$

or equivalently, we have

$$\int \frac{d W (\varphi_1, \varphi_2)}{W (\varphi_1, \varphi_2)} = -a_1 \int dx$$

this yields to

$$\ln W = -a_1 x + c$$

with c the integration constant. Finally, the solution of (9.32) is

$$W (\varphi_1, \varphi_2) = A e^{-a_1 x}$$

where $A = e^c$. ◻

Remark 9.2 If the roots of characteristic equation $\lambda^2 + a_1 \lambda + a_0 = 0$ of (9.29) $r_1, r_2 \in \mathbb{C}$ given by

$$r_1 = a + bi \quad, \quad r_2 = a - bi, \quad a, b \in \mathbb{R}, \ b \neq 0$$

then the general solution of differential equation (9.29) is

$$
\begin{aligned}
\varphi(x) &= c_1 e^{r_1 x} + c_2 e^{r_2 x} \\
&= c_1 e^{(a+bi)x} + c_2 e^{(a-bi)x} \\
&= e^{ax} \left[c_1 e^{ibx} + c_2 e^{-ibx} \right] \\
&= e^{ax} [c_1 (\cos bx + i \sin bx) + c_2 (\cos bx - i \sin bx)] \\
&= e^{ax} [(c_1 + c_2) \cos bx + (c_1 - c_2)i \sin bx] \\
&= A e^{ax} \cos bx + B e^{ax} \sin bx, \quad A = (c_1 + c_2), \ B = i(c_1 - c_2)
\end{aligned}
$$

In particular we have the following solutions

$$\varphi_1(x) = e^{ax} \cos bx \quad, \quad \varphi_2(x) = e^{ax} \sin bx$$

and $W (\varphi_1, \varphi_2) = b e^{2ax}$.

Example 9.13 Given the differential equation

$$y'' + k^2 y = \cos \omega x$$

we have $\lambda^2 + k^2 = 0$, whose roots are $\lambda_{1,2} = \pm ki$. According to the above remark,

$$\varphi_1(x) = e^{kxi} = \cos kx + i \sin kx$$
$$\varphi_2(x) = e^{-kxi} = \cos kx - i \sin kx$$

then
$$\varphi(x) = A \cos kx + B \sin kx, \quad A = c_1 + c_2, B = i(c_1 - c_2)$$

We define $\bar{\varphi}_1 = \cos kx$, $\bar{\varphi}_2 = \sin kx$ then $W(\bar{\varphi}_1, \bar{\varphi}_2) = k \neq 0$, that implies that $\mu_1' = -\dfrac{1}{k} \cos(\omega x) \sin(kx)$, $\mu_2' = \dfrac{1}{k} \cos(\omega x) \cos(kx)$. Finally the solution is given by $\varphi = \mu_1 A \bar{\varphi}_1 + \mu_2 B \bar{\varphi}_2$. It is left as an exercise to obtain μ_1, μ_2.

Definition 9.9 Let $\varphi_1(x)$, $\varphi_2(x)$ two solutions of the homogeneous equation

$$y'' + a_1 y' + a_0 y = 0$$

The solutions are said to be **linearly independents** (or independents) if

$$c_1 \varphi_1(x) + c_2 \varphi_2(x) = 0$$

then $c_1 = c_2 = 0$.

Proposition 9.3 *Two solutions $\varphi_1(x)$ and $\varphi_2(x)$ are linearly independents if and only if the Wronskian is not zero, i.e. $W(\varphi_1, \varphi_2) \neq 0$.*

Proof (Sufficiency) If $W(\varphi_1, \varphi_2) \neq 0$, i.e.,

$$W(\varphi_1, \varphi_2) = \begin{vmatrix} \varphi_1 & \varphi_2 \\ \varphi_1' & \varphi_2' \end{vmatrix} = \varphi_1 \varphi_2' - \varphi_2 \varphi_1' \neq 0$$

then

$$c_1 = \frac{\begin{vmatrix} 0 & \varphi_2 \\ 0 & \varphi_2' \end{vmatrix}}{W(\varphi_1, \varphi_2)} = \frac{0}{W(\varphi_1, \varphi_2)} = 0, \quad c_2 = \frac{\begin{vmatrix} \varphi_1 & 0 \\ \varphi_1' & 0 \end{vmatrix}}{W(\varphi_1, \varphi_2)} = \frac{0}{W(\varphi_1, \varphi_2)} = 0$$

Hence we have the trivial solution $c_1 = c_2 = 0$, thus φ_1 and φ_2 are linearly independents.

(Necessity) Suppose φ_1 and φ_2 are linearly independents. If $W(\varphi_1, \varphi_2) = 0$, we have hat the determinant is equal zero thus φ_1 and φ_2 are linearly dependents which is a contradiction. Hence $W(\varphi_1, \varphi_2) \neq 0$.   □

9.10 Indeterminate Coefficients

The method of indeterminate coefficients is used to find particularly solutions of the non-homogeneous linear equation of second order with constants coefficients. Before reviewing the methodology, some examples of motivation are presented.

Example 9.14 1. The function $y = 5$ is a particular solution of the differential equation $y'' + 2y' + y = 5$.
2. For the differential equation $y'' + y' - y = 3x$, the function $y = -3x - 3$ is a particular solution since $y' = -3$ and $y'' = 0$.
3. If $\varphi(x) = ax + b$ satisfies the differential equation $y'' + 5y' + 2y = 2x$, then $\varphi'(x) = a, \varphi''(x) = 0$ so that $5a + 2(ax + b) = 2x$. It is not hard to see that $a = 1, b = -\dfrac{5}{2}$. Therefore $\varphi(x) = x - 5/2$.
4. Let $y'' + 6y = 2x^2 + x - 3$ a given differential equation (DE). The function $\varphi(x) = ax^2 + bx + c$ is proposed as the solution for the DE. The problem consists in determine the values of a, b and c of $\varphi(x)$. Differentiating $\varphi(x)$ we have that $\varphi'(x) = 2ax + b, \varphi''(x) = 2a$ then substituting in the DE:

$$2a + 6(ax^2 + bx + c) = 2x^2 + x - 3$$

or equivalently

$$6ax^2 + 6bx + (2a + 6c) = 2x^2 + x - 3$$

It is said that two polynomials of degree n are equal if the coefficients of the monomials of equal degree are equal, then $6a = 2, 6b = 1$ and $2a + 6c = -3$, therefore

$$\varphi(x) = -\frac{11}{18}x^2 + \frac{1}{6}x + \frac{1}{3}$$

Exercise 9.5 For the differential equation $y'' + 2y' + y = \sin x$, the general form of the solution is given by $\varphi(x) = A \sin x + B \cos x$. According to the methodology of the examples, what are the values of A and B for the proposed solution $\varphi(x)$?

The examples provide information that leads to the general method to solve the following differential equation:

$$y'' + a_1 y' + a_0 y = b(x) \tag{9.33}$$

If $b(x) = a_n x^n + a_{n-1} x^{n-1} + \cdots + a_1 x + a_0$, then a polynomial of the same degree is proposed as the solution and the coefficients must be determined:

$$\varphi_p(x) = \alpha_n x^n + \alpha_{n-1} x^{n-1} + \cdots + \alpha_1 x + \alpha_0$$

On the other hand, if $b(x) = A \sin \omega x + B \cos \omega x$, the proposed solution is given by

$$\varphi_p(x) = A \sin \omega x + B \cos \omega x$$

even if any of the coefficients A or B are zero, i.e, $\varphi_p(x) = A \sin \omega x$ or $\varphi_p(x) = B \cos \omega x$.

In the same manner, if $b(x) = ae^{kx}$, then the proposed solution is given by $\varphi_p(x) = Ae^{kx}$. To determine the value of A, note that $\varphi'(x) = Ake^{kx}$, $\varphi''(x) = Ak^2e^{kx}$ then substituting in the DE:

$$Ak^2e^{kx} + a_1 Ake^{kx} + a_0 Ae^{kx} = ae^{kx}$$

so that:

$$A = \frac{a}{k^2 + a_1 k + a_0}$$

Remark 9.3 Consider the Eq. (9.33). To apply the method of indeterminate coefficients, $b(x)$ must satisfy the following. There exist $b_1, b_2, \ldots, b_n$ such that any derivative $b^{(k)}(x)$ must be able to be written as a linear combination of them. To verify this fact, it is necessary to derive $b(x)$ a sufficiently large number of times and observing that after a certain order of the derivative, all the terms that appear are repeated.

Exercise 9.6 Find a solution to the following differential equations.

1. $y'' + 3y = e^{2x} \sin x$. Hint: $\varphi(x) = Ae^{2x} \sin x + Be^{2x} \cos x$.
2. $y'' + 3y' = xe^x + x^2 \cos x$

9.11   Solution of Differential Equations by Means of Power Series

9.11.1   Some Criterions of Convergence of Series

Let a sequence $\{a_n\}_{n=1}^{\infty}$, the serie with general term a_n is the sequence $\{s_n\}_{n=1}^{\infty}$ defined by

$$s_1 = a_1$$
$$s_2 = a_1 + a_2$$
$$s_3 = a_1 + a_2 + a_3$$
$$\vdots$$
$$s_n = a_1 + a_2 + \cdots + a_n$$
$$s_n = \sum_{k=1}^{n} a_k$$

i.e.

$$\{s_n\}_{n=1}^{\infty} = \left\{\sum_{k=1}^{n} a_k\right\}_{n=1}^{\infty}$$

Remark 9.4 This sequence will be denoted by $\sum a_n$. If the index of a_n begins in n_0, i.e. $\{a_n\}_{n=n_0}^{\infty}$, then the sequence will be denoted by

$$\sum_{n=n_0} a_n$$

It is said that $\sum a_n$ **converges** if the sequence $s_1 = a_1, s_2 = a_1 + a_2, s_3 = a_1 + a_2 + a_3, \ldots$, $(s_1, s_2, s_3, \ldots$ are so called partial sums) converges. In this case, when a series $\sum a_n$ converges, the limit is denoted by

$$\sum_{n=1}^{\infty} a_n \quad \text{or} \quad \sum_{n=n_0}^{\infty} a_n$$

Definition 9.10 A sequence (s_n) is said to be a Cauchy sequence if for every positive real number ε, there exists a positive integer N such that for all natural numbers $m, n > N$

$$|s_n - s_m| < \varepsilon$$

This condition is usually written as $\lim_{m,n \to \infty} |s_m - s_n| = 0$.

Theorem 9.4 *A sequence (S_n) converges if and only if S_n is a Cauchy sequence.* □

Another form of the previous result is the following: A sequence (S_n) converges if and only if for every $\varepsilon > 0$ there exists a n_0 such that $n \geq n_0$ and $p \geq 1$:

$$|s_{n+p} - s_n| \leq \varepsilon$$

In particular if (S_n) is a $\sum a_n$ then the convergence criterion of a series is: $\sum a_n$ converges if and only if for every $\varepsilon > 0$ there exists a n_0 such that $n \geq n_0$ and $p \geq 1$ then:

$$\left|\sum_{k=1}^{p} a_{n+k}\right| < \varepsilon$$

Next the criterion for the convergence of a series is explained:

Theorem 9.5 D'Alembert criterion. *Let suppose a series $\sum a_n$, with $a_n \neq 0 \, \forall n$ such that exists the limit*

$$\lim_{n \to \infty} \frac{a_{n+1}}{a_n} = \alpha$$

then

1. *If $\alpha < 1$, the series converges.*
2. *If $\alpha > 1$, the series diverges.*
3. *None of the conclusions can be extended to the equality case $\alpha = 1$ (the series may converge or diverge).*   □

Example 9.15   1. Let the series $\displaystyle\sum_{n=1}^{\infty} \frac{1}{n}$, then

$$\alpha = \lim_{n \to \infty} \frac{\frac{1}{n+1}}{\frac{1}{n}} = \lim_{n \to \infty} \frac{n}{n+1} = 1$$

However the series does not converge for $\varepsilon = \dfrac{1}{10}$ with $N = N_0$, $P = 2N_0$ (see Cauchy criterion):

$$\frac{1}{10} > \sum_{n=N_0+1}^{2N_0} \frac{1}{n} = \frac{1}{N_0+1} + \frac{1}{N_0+2} + \cdots + \frac{1}{N_0+N_0} > \frac{N_0}{2N_0} = \frac{1}{2}$$

which is a contradiction, hence the series diverges.

2. Let the series $\displaystyle\sum_{n=1}^{\infty} \frac{1}{n(n+1)}$, then

$$\alpha = \lim_{n \to \infty} \frac{n}{n+2} = 1$$

On the other hand note that:

$$\sum_{n=1}^{\infty} \frac{1}{n(n+1)} = \lim_{n \to \infty} \sum_{k=1}^{n} \left(\frac{1}{k} - \frac{1}{k+1} \right)$$

$$= \lim_{n \to \infty} \left[\left(1 - \frac{1}{2} \right) + \left(\frac{1}{2} - \frac{1}{3} \right) + \left(\frac{1}{3} - \frac{1}{4} \right) + \cdots + \left(\frac{1}{n} - \frac{1}{n+1} \right) \right]$$

$$= \lim_{n \to \infty} \left(1 - \frac{1}{n+1} \right)$$

$$= 1$$

Hence, the series converges.

Theorem 9.6 *Let* $\{a_n\}$ *a sequence,* $x_0 \in \mathbb{R} \, \forall \, x \in \mathbb{R}$. *Let the power series* $\sum a_n (x - x_0)^n$, *then there exists* R, $0 \le R < \infty$ *such that the series* $\sum a_n (x - x_0)^n$:

1. *Converges if* $| \, x - x_0 \, | < R \, (x_0 - R < x < x_0 + R)$.
2. *Diverges if* $| \, x - x_0 \, | > R \, (x_0 - R > x > x_0 + R)$.
3. *If* $R = 0$, *the series converges only in* $x = x_0$.
4. *If* $R = \infty$, *the series converges for all* x. □

Remark 9.5 Using the Cauchy criterion, we have

$$\alpha = \lim_{n \to \infty} \left| \frac{a_{n+1}(x-x_0)^{n+1}}{a_n(x-x_0)^n} \right| = \lim_{n \to \infty} \left| \frac{a_{n+1}}{a_n} \right| | \, x - x_0 \, | = | \, x - x_0 \, | \underbrace{\lim_{n \to \infty} \left| \frac{a_{n+1}}{a_n} \right|}_{K} = | \, x - x_0 \, | \, K$$

The series converges if $| \, x - x_0 \, | \, K < 1$, that is, if $| \, x - x_0 \, | < \frac{1}{K} = R$. In other words, the series converges if $| \, x - x_0 \, | < R$, where

$$R = \frac{1}{\lim_{n \to \infty} \left| \frac{a_{n+1}}{a_n} \right|} = \lim_{n \to \infty} \left| \frac{a_n}{a_{n+1}} \right|$$

Example 9.16 1. The series $\sum \frac{1}{n!} x^n$ converges for all x. Indeed:

$$R = \lim_{n \to \infty} \frac{\frac{1}{n!} x^n}{\frac{1}{(n+1)!} x^{n+1}} = \lim_{n \to \infty} \frac{1}{x}(n+1) = \infty$$

2. The series $\sum x^n = 1 + x + x^2 + \cdots + x^n = \dfrac{1 - x^{n+1}}{1 - x^n}$ converges for all x, since

$$R = \lim_{n \to \infty} \left| \frac{a_n}{a_{n+1}} \right| = \lim_{n \to \infty} \left| \frac{x^n}{x^{n+1}} \right| = \lim_{n \to \infty} \left| \frac{1}{x} \right| = \infty \text{ if and only if } | \, x \, | < 1$$

9.11.2 Solution of First and Second Order Differential Equations

Example 9.17 Solve the following differential equations using power series.

1. Let $y' - y = 0$, then according to the main result of this section

$$y(x) = \sum_{n=0}^{\infty} a_n x^n$$

that converges for $|x| < R$, $R > 0$ and

$$y'(x) = \sum_{n=1}^{\infty} a_n n x^{n-1}$$

On the other hand, substituting y and y' in the differential equation:

$$y' - y = \sum_{n=1}^{\infty} a_n n x^{n-1} - \sum_{n=0}^{\infty} a_n x^n = \sum_{n=1}^{\infty} \left[a_n n - a_{n-1} \right] x^{n-1} = 0 \qquad (9.34)$$

Since (9.34) is equal to zero, then the coefficient of each power of x is equal to zero, that is to say,

$$a_n n - a_{n-1} = 0, \quad n = 1, 2, 3, \ldots \qquad (9.35)$$

The above recurrence relation determines a_n and since $n \neq 0$:

$$a_n = \frac{a_{n-1}}{n}, \quad n = 1, 2, 3, \ldots \qquad (9.36)$$

then:

$$a_1 = a_0$$
$$a_2 = \frac{a_1}{2} = \frac{1}{2!} a_0$$
$$a_3 = \frac{a_2}{3} = \frac{1}{3!} a_0$$
$$\vdots$$
$$a_n = \frac{1}{n!} a_0$$

Finally, substituting the obtained coefficients in the proposed solution:

$$y = \sum_{n=0}^{\infty} \frac{1}{n!} a_0 x^n = a_0 \sum_{n=0}^{\infty} \frac{x^n}{n!} = a_0 e^x$$

2. For the differential equation $(1 + x)y' = Py$, $y(0) = 1$, let

$$y(x) = \sum_{n=0}^{\infty} a_n x^n$$

that converges for $|x| < R$, $R > 0$ and

$$y'(x) = \sum_{n=1}^{\infty} a_n n x^{n-1}$$

Substituting in the differential equation we have:

$$(1+x)y' = Py$$

$$(1+x) \sum_{n=1}^{\infty} a_n n x^{n-1} = P \sum_{n=0}^{\infty} a_n x^n$$

$$\sum_{n=1}^{\infty} a_n n (1+x) x^{n-1} = \sum_{n=0}^{\infty} P a_n x^n$$

$$\sum_{n=1}^{\infty} a_n n (x^{n-1} + x^n) = \sum_{n=0}^{\infty} P a_n x^n$$

$$\sum_{n=1}^{\infty} a_n n x^{n-1} + \sum_{n=1}^{\infty} a_n n x^n = P a_0 + \sum_{n=1}^{\infty} P a_n x^n$$

$$a_1 + \sum_{n=1}^{\infty} \left[a_{n+1}(n+1) + a_n n \right] x^n = P a_0 + \sum_{n=1}^{\infty} P a_n x^n$$

then $a_1 = a_0 P$ (coefficients of x^0 in both sides) and $a_{n+1}(n+1) + a_n n = P a_n$ for $n \in \mathbb{N}$. Since $y(0) = 1$, according to the proposed solution, $a_1 = P$. The other coefficients are determined with the recurrence relation:

$$a_{n+1} = \frac{a_n (P - n)}{n + 1}, \quad n = 1, 2, 3, \ldots \tag{9.37}$$

For some positive integers, we have:

$$a_2 = \frac{a_1(P-1)}{2} = \frac{(P-1)P}{2}, \quad n = 1$$

$$a_3 = \frac{a_2(P-2)}{3} = \frac{P-2}{3} \cdot \frac{(P-1)P}{2}, \quad n = 2$$

$$a_4 = \frac{a_3(P-3)}{4} = \frac{P-3}{4} \cdot \frac{P-2}{3} \cdot \frac{(P-1)P}{2}, \quad n = 3$$

$$\vdots$$

$$a_{k+1} = \frac{a_k(P-k)}{k+1} = \frac{P-k}{k+1} \cdot \frac{P-(k-1)}{k} \cdot \frac{P-(k-2)}{k-1} \cdots P = \frac{P(P-1)(P-2) \cdots (P-k)}{(k+1)!}$$

then

$$a_n = \frac{(n+1)}{P-n}a_{n+1}$$
$$= \frac{(n+1)}{P-n}\frac{P(P-1)(P-2)\cdots(P-n)}{(n+1)!}$$
$$= \frac{P(P-1)(P-2)\cdots(P-n+1)}{n!}$$
$$= \frac{P!}{n!(P-n)!}$$
$$a_n = \binom{P}{n}$$

hence

$$y = \sum_{n=0}^{\infty} \binom{P}{n} x^n$$

$$y'' - 2xy' - 2y = 0, \quad y(0) = a_0, \quad y'(0) = a_1 \tag{9.38}$$

Let

$$y(x) = a_0 + a_1 x + a_2 x^2 + \cdots = \sum_{n=0}^{\infty} a_n x^n$$

and its derivatives sequential given by

$$y'(x) = a_1 + 2a_2 x + 3a_3 x^2 + \cdots + = \sum_{n=1}^{\infty} n a_n x^{n-1}$$

$$y''(x) = 2a_2 + 6a_3 x + \cdots + = \sum_{n=2}^{\infty} n(n-1) a_n x^{n-2}$$

Substituting in (9.38):

$$\sum_{n=2}^{\infty} n(n-1)a_n x^{n-2} - 2x\sum_{n=1}^{\infty} n a_n x^{n-1} - 2\sum_{n=0}^{\infty} a_n x^n$$

Exercise 9.7 Solve the following differential equations:

1. $y' = 2xy$
2. $y' + y = 1$

Example 9.18 Consider the differential equation of second order:

$$y'' - 2xy' - 2y = 0, \quad y(0) = a_0, \quad y'(0) = a_1 \tag{9.39}$$

Let

$$y(x) = a_0 + a_1 x + a_2 x^2 + \cdots = \sum_{n=0}^{\infty} a_n x^n$$

whose first and second derivatives are given by

$$y'(x) = a_1 + 2a_2 x + 3a_3 x^2 + \cdots = \sum_{n=1}^{\infty} n a_n x^{n-1}$$

$$y''(x) = 2a_2 + 6a_3 x + \cdots = \sum_{n=2}^{\infty} n(n-1) a_n x^{n-2}$$

Substituting the previous expressions in the Eq. (9.39):

$$\sum_{n=2}^{\infty} n(n-1) a_n x^{n-2} - 2x \sum_{n=1}^{\infty} n a_n x^{n-1} - 2 \sum_{n=0}^{\infty} a_n x^n = 0$$

since

$$\sum_{n=2}^{\infty} n(n-1) a_n x^{n-2} = \sum_{n=0}^{\infty} (n+2)(n+1) a_{n+2} x^n$$

this yields to

$$\sum_{n=0}^{\infty} (n+2)(n+1) a_{n+2} x^n - 2 \sum_{n=0}^{\infty} n a_n x^n - 2 \sum_{n=0}^{\infty} a_n x^n = 0.$$

We obtain $(n+2)(n+1) a_{n+2} - 2n a_n - 2a_n = 0$, so that

$$a_{n+2} = \frac{2a_n}{n+2}.$$

To find two solutions of Eq. (9.39), two different values for a_0 and a_1 are chosen. For example, $a_0 = 1$, $a_1 = 0$ or $a_0 = 0$, $a_1 = 1$. For the first case, the odd coefficients a_1, a_3, a_5 are zero because

$$a_3 = \frac{2a_1}{1+2} = 0, \ a_5 = \frac{2a_3}{5} = 0, \ldots$$

On the other hand, the even coefficients are determined by

$$a_2 = a_0 = 1, \ a_4 = \frac{2a_2}{4} = \frac{1}{2}, \ a_6 = \frac{2a_4}{6} = \frac{1}{2 \cdot 3}, \ldots, \ a_{2n} = \frac{1}{n!}$$

then one solution is:

$$y_1(x) = 1 + x + x^2 + \frac{x^4}{2!} + \frac{x^6}{3!} + \cdots = e^{x^2}.$$

For the second case, $a_0 = 0$, $a_1 = 1$, the even coefficients are zero and the odd coefficients are given by:

$$a_3 = \frac{2a_1}{3}, \; a_5 = \frac{2a_3}{5} = \frac{2}{5}\frac{2}{3}, \; a_7 = \frac{2a_5}{7} = \frac{2}{7}\frac{2}{5}\frac{2}{3}, \ldots, \; a_{2n+1} = \frac{2^n}{3 \cdot 5 \cdot 7 \cdots (2n+1)}$$

then, other solution is

$$y_2(x) = x + \frac{2}{3}x^3 + \frac{2^2}{3 \cdot 5}x^5 + \cdots = \sum_{n=0}^{\infty} \frac{2^n x^{2n+1}}{3 \cdot 5 \cdot 7 \cdots (2n+1)}$$

Exercise 9.8 Solve the following differential equations:

1. $y'' + y = 0$
2. $y' = x - y, \; y(0) = 0$

9.12 Picard's Method

Let the initial value problem

$$y' = f(t, y) \tag{9.40}$$

then, from (9.40) can be written in a special form, integrating both sides of equation respect to t. Namely, if $y(t)$ satisfies (9.40), we have:

$$\int_{t_0}^{t} \frac{dy(s)}{ds} = \int_{t_0}^{t} f(s, y(s))$$

then

$$y(t) = y_0 + \int_{t_0}^{t} f(s, y(s)) \, ds \tag{9.41}$$

Reciprocally, if $y(t)$ is a continuous function and satisfies (9.41), then $\frac{dy}{dt} = f(t, y(t))$. Therefore, $y(t)$ is a solution of (9.40), if and only if, it is a continuous solution of (9.41). This suggests the following method to obtain a succession of *approximate solutions* $y_n(t)$ from (9.41). The Eq. (9.41) is an integral equation and it is in the special form:

$$y_0 + \int_{t_0}^{t} f(s, y(s)) \, ds$$

The method begins by proposing a tentative solution $y_0(t)$ of (9.41). The easiest choice is $y_0(t) = y_0$. To check if $y_0(t)$ is a solution of (9.41), is calculated

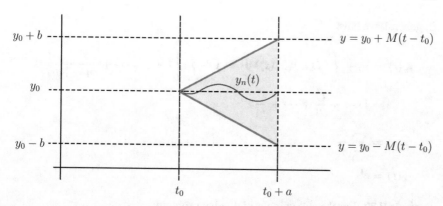

Fig. 9.4 Rectangular domain

$$y_1(t) = y_0 + \int_{t_0}^{t} f(s, y_0(s))\, ds$$

Now, the $y_1(t)$ as the next option is used. If $y_1(t) = y_0$, then $y(t) = y_0$. Otherwise, y_1 as the following approximation is used. To verify that $y_1(t)$ is a solution of (9.41), it is calculated

$$y_2(t) = y_0 + \int_{t_0}^{t} f(s, y_1(s))\, ds$$

In this form, we can define a sequence of functions $y_1, y_2, \ldots$, where

$$y_{n+1}(t) = y_0 + \int_{t_0}^{t} f(s, y_n(s))\, ds$$

The functions $y_n(t)$ are called *successive approximations* or *Picard's iterations* (Fig. 9.4).

Example 9.19 To calculate the Picard's iterations for the following initial value problem and verify that the sequences converge. Let

$$y' = y = f(t, y(t))\quad , y(0) = 1$$

The integral equation is

$$y_1(t) = y_0 + \int_{t_0}^{t} f(s, y_0(s))\, ds = 1 + \int_{0}^{t} 1\, ds = 1 + t$$

so that

$$y_2(t) = y_0 + \int_{0}^{t} f(s, y_1(s))\, ds = 1 + \int_{0}^{t} (1 + s)\, ds = 1 + t + \tfrac{t^2}{2}$$

In general, we have

$$y_n(t) = y_0 + \int_0^t f(s, y_{n-1}(s)) \, ds = 1 + \int_0^t \left[1 + s + \cdots + \frac{s^{n-1}}{(n-1)!}\right] ds$$

$$= 1 + t + \frac{t^2}{2!} + \cdots + \frac{t^n}{n!}$$

$$= \sum_{n=0}^{\infty} \frac{t^n}{n!}$$

$$y_n(t) = e^t$$

Example 9.20 For the following initial value problem,

$$y' = 1 + y^3, \quad y_0(1) = 1$$

the corresponding integral equations is:

$$y(t) = 1 + \int_1^t (1 + y^3(s)) ds$$

For $y_0(t) = 1$, we have $f(1, y_0(1)) = 1 + 1^3 = 2$, then:

$$y_1(t) = 1 + \int_1^t 2ds = 1 + 2(t - 1)$$

and for $y_2(t)$, in the same manner:

$$y_2(t) = 1 + \int_1^t \left[1 + 2(s - 1)^3\right] ds$$

$$= 1 + \left[2s + 3(s - 1)^2 + 4(s - 1)^3 + 2(s - 1)^4\right]_1^t$$

$$= 1 + 2(t - 1) + 3(t - 1)^2 + 4(t - 1)^3$$

$$y_2(t) = \sum_{k=1}^{\infty} K (t - 1)^{K-1}$$

9.13 Convergence of Picard's Iterations

Lemma 9.1 *Let a, b positive numbers, let the rectangle R: $t_0 \leq t \leq t_0 + a$, $|y - y_0| \leq b$. Let $M = \max|f(t, y)|$ where $(t, y) \in R$ and $\alpha = \min\left(a, \frac{b}{M}\right)$ then $|y_n(t) - y_0| \leq M(t - t_0)$ for $t_0 \leq t \leq t_0 + \alpha$.*

Proof (From Lemma 9.1). It is fulfilled for $n = 0$, $y_0(t) = y_0$. For $n = j$:

$$|y_{j+1}(t) - y_0| = \left| \int_{t_0}^{t} f|s, y_j(s)|ds \leq \int_{t_0}^{t} |f(s, y_j)|ds \right| \leq M(t - t_0)$$

where

$$M = \max_{(t,y)\in R} |f(t, y)|$$

Now it is possible to prove that Picard's iterations $y_n(t)$ converge for all t in the interval $t_0 \leq t \leq t_0 + \alpha$. If $\frac{\partial f}{\partial y}$ exists and it is continuous:

$$y_n(t) = y_0(t) + [y_1(t) - y_0] + [y_2 - y_1] + [y_3 - y_2] + \cdots + \left[y_n - y_{n-1}\right]$$

It is clear that sequence $y_n(t)$ converges if and only if the series converge:

$$|y_n - y_0| = |y_1 - y_0(t) - \ldots - y_n(t) - y_{n-1}(t)|$$

that is:

$$\sum_{n=1}^{\infty} |y_n(t) - y_{n-1}(t)| < \infty$$

Note that

$$|y_n(t) - y_{n-1}(t)| = \left| \int_{t_0}^{t} (f(s, y_{n-1}(s)) - f(s, y_{n-2}(s))) \, ds \right| \leq \int_{t_0}^{t} |f(s, y_{n-1}(s)) - f(s, y_{n-2}(s))|ds$$

$$= \int_{t_0}^{t} |\frac{\partial f(s, \zeta(s))}{\partial y}||y_{n-1}(s) - y_{n-2}(s)|ds$$

where $\zeta(s)$ is between $y_{n-1}(s)$ and $y_{n-2}(s)$. Therefore

$$|y_n(t) - y_{n-1}(t)| \leq L \int_{t_0}^{t} |y_{n-1}(s) - y_{n-2}(s)|ds$$

for $t_0 \leq t \leq t_0 + \alpha$ and

$$L = \max_{(t,y)\in R} |\frac{\partial f(t,y)}{\partial y}| \qquad\qquad \square$$

Remark 9.6 If $y \leq y_0 + b$ then $y_0 + M(t - t_0) \leq y_0 + b$. We obtain $a \leq \frac{b}{M}$.

Definition 9.11 A function f(x,y) defined in a domain D it is said to satisfy the Lipschitz condition with respect to y if for every x, y_1, y_2 such that (x, y_1), $(x, y_2) \in D$:

$$|f(x, y_1) - f(x, y_2)| \leq K|y_1 - y_2|$$

where K is called the Lipschitz constant.

Lemma 9.2 *Let $f(x, y)$ with partial derivative $\frac{\partial f}{\partial y}(x, y)$ bounded for every $(x, y) \in D$, D is a convex set, then $f(x, y)$ satisfies the Lipschitz condition with constant K such that*

$$\left| \frac{\partial f(x, y)}{\partial y} \right| \leq K, \quad (x, y) \in D$$

Proof Using the Mean Value Theorem:

$$|f(x, y_2) - f(x, y_1)| = \left| \frac{\partial f(x, \xi)}{\partial y}(y_2 - y_1) \right| \leq \left| \frac{\partial f(x, \xi)}{\partial y} \right| |y_2 - y_1| \leq K |y_2 - y_1|$$

where

$$K = \max_{(x, \xi) \in D} \left| \frac{\partial f(x, \xi)}{\partial y} \right| \qquad \square$$

Theorem 9.7 (Theorem of approximated solutions) *Let $(x_0, y_0) \in R$, $f(x, y)$ continuous function that satisfies the Lipschitz conditions with constant K. Let $y_1(x)$, $y_2(x)$ approximated solutions of the differential equation $y' = f(x, y)$ defined for $|x - x_0| \leq h$ with errors $\varepsilon_1, \varepsilon_2$. Define $P(x) = y_1(x) - y_2(x)$ and $\varepsilon = \varepsilon_1 + \varepsilon_2$, then:*

$$|P(x)| \leq e^{K|x-x_0|} |P(x_0)| + \frac{\varepsilon}{K} \left(e^{K|x-x_0|} - 1 \right)$$

Proof The inequality $|x - x_0| \leq h$ defines the interval $[x_0 - h, x_0 + h]$ (Without lost of generality, the proof is only for the interval $x_0 \leq x \leq x_0 + h$).

Since the solutions $y_1(x)$ and $y_2(x)$ satisfy:

$$|\frac{dy_1}{dx} - f(x, y_1)| \leq \varepsilon_1, \quad |\frac{dy_2}{dx} - f(x, y_2)| \leq \varepsilon_2$$

then

$$\left| \frac{dy_1}{dx} - \frac{dy_2}{dx} \right| = |\frac{dy_1}{dx} - \frac{dy_2}{dx} + f(x, y_1) - f(x, y_1) + f(x, y_2) - f(x, y_2)|$$

$$= \left| \left[\frac{dy_1}{dx} - f(x, y_1) \right] - \left[\frac{dy_2}{dx} + f(x, y_2) \right] + [f(x, y_1) - f(x, y_2)] \right|$$

$$\leq \varepsilon_1 + \varepsilon_2 + |f(x, y_1) - f(x, y_2)|$$

$$\leq \varepsilon_1 + \varepsilon_2 + K |y_1 - y_2|$$

$$= \varepsilon + K |y_1 - y_2|$$

On the other hand, since $P(x) = y_1(x) - y_2(x)$ then $P'(x) = \frac{dy_1}{dx} + \frac{dy_2}{dx}$, so that

$$|P'(x)| \leq \varepsilon + K |P(x)|$$

Now consider the following three cases:

1. $P(x) \neq 0$, $x_0 \leq x \leq x_0 + h$. For $P(x) > 0$:

$$\frac{dP(x)}{dx} \leq \left| \frac{dP(x)}{dx} \right| \leq KP(x) + \varepsilon$$

$$P'(x) \leq KP(x) + \varepsilon$$

$$P'(x) - KP(x) \leq \varepsilon$$

$$e^{-Kx}\left[P'(x) - KP(x) \right] \leq e^{-Kx}\varepsilon$$

$$\left[e^{-Kx}P(x) \right]' \leq e^{-Kx}\varepsilon$$

Integrating both sides:

$$\int_{x_0}^{x} \left[e^{-Kx}P(x) \right]' dx \leq \int_{x_0}^{x} e^{-Kx}\varepsilon dx$$

$$e^{-Kx}P(x)\Big|_{x_0}^{x} \leq \frac{\varepsilon}{K}\left[e^{-Kx_0} - e^{-Kx} \right]$$

$$e^{-Kx}P(x) - e^{-Kx_0}P(x_0) \leq \frac{\varepsilon}{K}\left[e^{-Kx_0} - e^{-Kx} \right]$$

$$P(x) \leq e^{K(x-x_0)}P(x_0) + \frac{\varepsilon}{K}\left[e^{K(x-x_0)} - 1 \right]$$

2. For $P(x) = 0$. In this case $|P(x)| = 0$. This value is bounded for every positive constant.
3. $P(x) < 0$. The proof is left to the reader as an exercise. □

If D is a connected[1] region in $\mathbb{R}$, we have the following existence theorem for solutions of linear differential equations.

Theorem 9.8 *Let $D \subset \mathbb{R}$ be simply connected and B and C holomorphic[2] on D. Let a homogeneous linear differential equation:*

$$\frac{d^2x}{dz^2} + B(z)\frac{dx}{dz} + C(z)x = 0 \tag{9.42}$$

with initial conditions at $z = z_0 \in D$ given by

$$x(z_0) = \theta$$

$$\frac{dx}{dz}(z_0) = \varepsilon \tag{9.43}$$

---

[1]Roughly speaking, a region D is connected if for an arbitrary pair of points P and Q, we can always draw a curve in D.

[2]A function F defined on D (open, connected subset) is holomorphic if $F(\cdot)$ is differentiable and $F'(\cdot)$ is continuous.

has a unique holomorphic solution x on D satisfying both (9.42) *and* (9.43).

The proof can be found in [8]. Since (9.42) is a homogeneous equation, we state that the set V of functions that satisfy the above differential equation is a vector space over $\mathbb{C}$. Let $x_1, \ldots, x_n \in V$ and $\lambda_1, \ldots, \lambda_n \in \mathbb{C}$ then we have $\sum_{i=1}^{n} \lambda_i x_i \in V$. To see this, note that

$$x_i \in V \quad \text{if and only if} \quad \frac{d^2 x_i}{dz^2} + B(z)\frac{dx_i}{dz} + C(z)x_i = 0$$

Multiplying by λ_i and add, then

$$\frac{d^2}{dz^2}\left(\sum_{i=1}^{n} \lambda_i x_i\right) + B(z)\frac{d}{dz}\left(\sum_{i=1}^{n} \lambda_i x_i\right) + C(z)\left(\sum_{i=1}^{n} \lambda_i x_i\right) = 0$$

Thus, $\sum_{i=1}^{n} \lambda_i x_i$ is also a solution of (9.42). Since the differential equation (9.42) has order 2, the dimension of V over $\mathbb{C}$ is 2, i.e., dim $V = 2$. This can be see as follows. We define a map:

$$\varphi : V \longrightarrow \mathbb{C}^2$$

$$x \mapsto \varphi(x) = \begin{pmatrix} x(z_0) = \theta \\ \frac{dx}{dz}(z_0) = \varepsilon \end{pmatrix}$$

Clearly φ is a linear map from V to $\mathbb{C}^2$. By the existence theorem above, if $(\theta, \varepsilon)^\mathsf{T} \in \mathbb{C}^2$ is an arbitrary vector, there is a unique solution $x \in V$ such that

$$\begin{pmatrix} x(z_0) \\ \frac{dx}{dz}(z_0) \end{pmatrix} = \begin{pmatrix} \theta \\ \varepsilon \end{pmatrix}$$

Therefore, φ is bijective. Thus $V \cong \mathbb{C}^2$ (V is isomorphic to $\mathbb{C}^2$) and dim $V = 2$. $\square$

Example 9.21 An interesting problem in control theory, especially in linear time invariant systems the state-model takes the following form (see Chap. 7, Theorem 7.2):

$$\dot{x} = Ax(t) + Bu(t) \tag{9.44}$$

that represents an nonhomogeneous differential equation. We obtain the matrix solution of (9.44) applying the integrating factor method. The integrating factor is chosen as e^{-At}. Multiplying both sides of (9.44) by the integrating factor and rearranging terms:

$$\left(e^{-At}\right)\dot{x} = \left(e^{-At}\right)(Ax(t) + Bu(t))$$
$$\dot{x}e^{-At} = Ax(t)e^{-At} + Bu(t)e^{-At}$$
$$\dot{x}e^{-At} - Ax(t)e^{-At} = Bu(t)e^{-At}$$

Note that $\frac{d}{dt}\left(e^{-At}x(t)\right) = \dot{x}e^{-At} - Ax(t)e^{-At}$, then:

$$\frac{d}{dt}\left(e^{-At}x(t)\right) = e^{-At}Bu(t) \qquad (9.45)$$

Integrating both sides of above equation:

$$\int_0^t d\left(e^{-A\tau}x(\tau)\right) = \int_0^t e^{-A\tau}Bu(\tau)d\tau$$

$$e^{-A\tau}x(\tau)\Big|_0^t = \int_0^t e^{-A\tau}Bu(\tau)d\tau$$

$$e^{-At}x(t) - e^0 x(0) = \int_0^t e^{-A\tau}Bu(\tau)d\tau$$

$$e^{-At}x(t) = x(0) + \int_0^t e^{-A\tau}Bu(\tau)d\tau$$

since $e^0 = I$. Finally, multiplying both sides by the inverse of e^{-At}, i.e. e^{At} the solution of (9.44) is given by:

$$x(t) = x_0 e^{At} + \int_0^t e^{A(t-\tau)}Bu(\tau)d\tau \qquad (9.46)$$

where $x_0 = x(0)$.

References

1. Boyce, W. E., DiPrima, R. C.: Differential Equations Elementary and Boundary Value Problems. Willey (1977)
2. Walter, W.: Ordinary Differential Equations. Springer, New York (1998)
3. Murphy, G.M.: Ordinary Differential Equations and their Solutions. Dover Publications, New York (2011)
4. Simmons, G.F.: Differential Equations With Applications and Historical Notes. CRC Press (2016)
5. Hartman, P.: Ordinary Differential Equations. Willey (1964)
6. Hale, J.: Ordinary Differential Equations. Dover Publications, New York (1969)
7. Tricomi, F.G.: Differential Equations. Dover Publications, New York (2012)
8. Birkhoff, G., Rota, G.C.: Ordinary Differential Equations. Wiley, New York (1989)

Chapter 10
Differential Algebra for Nonlinear Control Theory

Abstract This chapter focus is to show the application of the commutative algebra, algebraic geometry and differential algebra concepts to nonlinear control theory, it begins with necessary information of differential algebra, it continues with definitions of single-input single-output systems, invertible systems, realization and canonical forms, finally we present methods for stabilization of nonlinear systems throughout linearization by dynamical feedback and some examples of such processes.

10.1 Algebraic and Transcendental Field Extensions

The notion of field was introduced in the late nineteenth century in Germany, with the purpose of treating equations while avoiding the need of complex calculations, in a way similar to nonlinear equations.

Definition 10.1 Let L, K fields such that $K \subset L$. The larger field L is then said to be an extension field of K denoted by L/K [1–4].

On the other hand, consider $a \in L$.

1. a is **algebraic** over K if there exists a polynomial $P(x)$ with coefficients in K such that $P(a) = 0$.
2. a is **transcendent** over K if it is not algebraic, that is, if there is no polynomial $P(x)$ with coefficients in K such that $P(a) = 0$.

For example, if $K = \mathbb{Q}$, $\sqrt{2} \in \mathbb{R}$ is algebraic over $\mathbb{Q}$ since there exists a polynomial, namely $p(x) = x^2 - 2$, such that $p(\sqrt{2}) = 0$. If $L = \mathbb{R}$, π, e are transcendental elements over $\mathbb{Q}$.

According to the above concepts, the extension L/K is algebraic if and only if every element of L is algebraic over K, it is said to be transcendental if and only if there exist at least one element of L such that it is transcendental over K.

© Springer Nature Switzerland AG 2019
R. Martínez-Guerra et al., *Algebraic and Differential Methods for Nonlinear Control Theory*, Mathematical and Analytical Techniques with Applications to Engineering, https://doi.org/10.1007/978-3-030-12025-2_10

Now, consider the algebraic dependence. This idea is similar to linear independence of vectors in vector spaces.

Definition 10.2 The elements $\{a_i \mid i \in I\}$ of L are said to be K-algebraically independent if there are no polynomials $P(x_1, \ldots, x_\gamma)$ in any finite number of variables with coefficients in K such that $P(a_1, \ldots, a_\gamma) = 0$. Elements no K-algebraically independent are said to be K-algebraically dependent [5–7].

It is possible to show the existence of maximal families of elements of L that are K-algebraically independent. Two maximal families have the same cardinality (the same number of elements).

Definition 10.3 Let S be a subset of L a maximal family. S is said to be a transcendence basis of L/K if it is algebraically independent over K and if furthermore L is an algebraic extension of the field K(S) (the field obtained by adjoining the elements of S to K). The cardinality of this basis is the transcendence degree of L/K denoted by $\operatorname{tr} d° L/K$ [8–10].

Example 10.1   1. The extension L/K is algebraic if and only if $\operatorname{tr} d° L/K = 0$.
2. Let $K = \mathbb{R}$. We consider the field of rational functions in the variable s, denoted by $\mathbb{R}(s)$:
$$f(s) = \frac{a_0 + a_1 s + \cdots + a_n s^n}{b_0 + b_1 s + \cdots + b_m s^m}, \quad a_i, b_i \in \mathbb{R}$$

Hence s is transcendental over $K = \mathbb{R}$, then $\operatorname{tr} d° L/K = 1$.
3. Let $K = \mathbb{R}$. The field of rational fractions in a finite number α of variables $s_1, \ldots, s_\alpha$:
$$f(s_1, \ldots, s_\alpha) = \frac{P(s_1, \ldots, s_\alpha)}{Q(s_1, \ldots, s_\alpha)}$$

P, Q are polynomials with real coefficients in $s_1, \ldots, s_\alpha$. This field is denoted by $\mathbb{R}(s_1, \ldots, s_\alpha)$. In this case, $\operatorname{tr} d° \mathbb{R}(s_1, \ldots, s_\alpha)/\mathbb{R} = \alpha$. $s_1, \ldots, s_\alpha$ are $\mathbb{R}$-algebraically independent (they constitute a transcendence basis). These extensions are called pure transcendental extensions.

Theorem 10.1 ([4]) *Let K, L, M be fields such that $K \subset L \subset M$ to be a tower of fields. Then*
$$\operatorname{tr} d° M/K = \operatorname{tr} d° M/L + \operatorname{tr} d° L/K$$
□

Example 10.2 As an application of the above Theorem, let L/K be a field extension of transcendence degree n, i.e., $\operatorname{tr} d° L/K = n$. Let $(x_1, \ldots, x_n)$ a transcendence basis of L/K. Considering the field $K(x_1, \ldots, x_n)$ we have $K \subset K(x_1, \ldots, x_n) \subset L$. By the Theorem:
$$n = \operatorname{tr} d° L/K = \operatorname{tr} d° L/K(x_1, \ldots, x_n) + \operatorname{tr} d° K(x_1, \ldots, x_n)/K$$

According to the last part of the above example, $\operatorname{tr} d^\circ K (x_1, \ldots, x_n) / K = n$, then $\operatorname{tr} d^\circ L/K (x_1, \ldots, x_n) = 0$. That means that algebraic extension $L/K (x_1, \ldots, x_n)$ is algebraic.

10.2 Basic Notions of Differential Algebra

The differential algebra was introduced by the American mathematician Joseph Ritt, It was introduced to provide the theory of differential equations tools similar to those given by commutative algebra for algebraic equations.

Definition 10.4 Let A a commutative ring with unity. A derivation of A is an application δ of A in A such that

$$\delta(a + b) = \delta(a) + \delta(b)$$
$$\delta(ab) = a\delta(b) + \delta(a)b$$

for all $a, b \in A$.

Definition 10.5 A differential ring A is a commutative ring with unity provided with a finite set Δ of derivations such that for all $\delta_1, \delta_2 \in \Delta$, for all $a \in A$, $\delta_1\delta_2 a = \delta_2\delta_1 a$.

Definition 10.6 A differential ring is said to be ordinary or partial if the finite set Δ contains one or more derivations, respectively.

Definition 10.7 A differential field is a differential ring.

Definition 10.8 A constant of a differential ring is an element a such that $\delta(a) = 0$ for all $\delta \in \Delta$.

Lemma 10.1 *The set of constants of a differential ring (respectively a differential field) is a differential subring (respectively a differential subfield).*

Proof Let A be a differential ring. Let C the set of constants. C contains at least 0 and 1.

1. For all $a \in A$, $a + 0 = a$, then $\delta(a + 0) = \delta(a) + \delta(0) = \delta(a)$ so that $\delta(0) = 0$.
2. In the same manner, $\delta(a) = \delta(a \cdot 1) = 1 \cdot \delta(a) + a \cdot \delta(1) = \delta(a) + a\delta(1)$ then $\delta(1) = 0$.
3. On the other hand, let $c_1, c_2 \in C$, then $\delta(c_1 + c_2) = \delta(c_1) + \delta(c_2) = 0$ and $\delta(c_1 c_2) = c_1\delta(c_2) + \delta(c_1)c_2 = 0$.
4. Finally, let c a nonzero constant, then $\frac{1}{c}$ is a constant. Besides for all $\delta \in \Delta$, $\delta\left(\frac{1}{c}\right) = -\frac{\delta(c)}{c^2} = 0$. $\qquad\square$

Definition 10.9 A differential field extension L/K is given by two differential fields K and L, such that:

1. K is subfield of L, $(K \subset L)$.
2. The derivation of K is the restriction to K of the derivation of L.

Example 10.3 $\mathbb{R}, \mathbb{Q}, \mathbb{C}$ are trivial fields of constants.

Let L/K a differential field extension. There are two possible cases:

1. An element $a \in L$ is said to be differentially algebraic over K if and only if there exists a differential algebraic extension with coefficients in K, i.e., there exist a polynomial $P(x, \dot{x}, \ldots, x^{(n)})$ with coefficients in K in a finite number of derivations of x such that $P(a, \dot{a}, \ldots, a^{(n)}) = 0$. The extension L/K is said to be differentially algebraic if and only if every element of L is differentially algebraic over K. (It is also called Hyperalgebraic).
2. An element $a \in L$ is said to be differentially transcendental over K if and only it is not differentially algebraic. The extension L/K is said to be differentially transcendental if and only if there exist at least an element of L differentially transcendental over K.

Example 10.4 $\mathbb{R}\langle e^t \rangle / \mathbb{R}$ is a differential field extension $\mathbb{R} \subset \mathbb{R}\langle e^t \rangle$ where e^t is a solution of $P(x) = \dot{x} - x = 0$ and it is differentially algebraic.

Definition 10.10 A set $\{\xi_i \mid i \in I\}$ of elements in L is said to be differentially K-algebraically dependent if and only if the set of derivatives of any order $\{\xi_i^{(v_i)} \mid i \in I, v_i = 0, 1, 2, \ldots\}$ is K-algebraically dependent.

Definition 10.11 If a set is not differentially K-algebraically dependent is said to be differentially K-algebraically independent [1, 3, 5].

Definition 10.12 Every family differentially K-algebraically independent such that is maximal respect to the inclusion is said to be a differential transcendence base of L/K. The cardinality of a a base is called the differential transcendence degree of L/K denoted by diff tr d° L/K.

Example 10.5 L/K is differentially algebraic if and only diff tr d° $L/K = 0$.

Theorem 10.2 *Let K, L, M fields, such that $K \subset L \subset M$ is a tower of differential field extensions. Then*

$$\text{diff tr d}° \, M/K = \text{diff tr d}° \, M/L + \text{diff tr d}° \, L/K$$

□

10.3 Definitions and Properties of Single-Input Single-Output Systems

One of the most well known topics on automatic control theory is stationary linear systems that posses a single-input and a single-output (SISO), usually, these system are described by a transfer function, further developments have made possible to include optimal control and optimal statistical filtering which led to the notion of state space representation for linear systems. these kind of systems can be better described by using algebra concepts as follows.

Let K be a differential field of a given basis. If the coefficients of the equations are constants then we take $K = \mathbb{R}$. On the other hand if the coefficients are variables, we take K as a field of meromorphic functions of single or several variables. Let $u = (u_1, \ldots, u_m)$ a finite family of differential quantities. $K \langle u \rangle$ denotes the differential field of K and the components of u.

Example 10.6

$$\frac{\left(u_1^{(28)}\right)^{45} \dot{u}_2^2 - 1}{\ddot{u}_1 + 3 \left(u_2^{(8)}\right)^6}$$

Let $u = (u_1, \ldots, u_m)$ and $y = (y_1, \ldots, y_n)$ two finite families of differential quantities. We denote by $K \langle u, y \rangle$ the differential field by K and the components of u and y.

Definition 10.13 A system withinput u and output y consists in a differential algebraic extension $K \langle u, y \rangle / K \langle u \rangle$, i.e., the components of $y_1, \ldots, y_n$ of y are differentially algebraic over $K \langle u \rangle$ [3, 6, 7].

Remark 10.1 In the definition of differential algebraic extensions, the differential algebraic equations are used. In the case of a simple pendulum, $\ddot{y} + \sin y = 0$, note the following situation. $\sin y$ satisfies a polynomial differential equation:

$$\frac{d}{dt} \sin y = \dot{y} \cos y$$

besides

$$\dot{y}^2 (\sin y)^2 + \left(\frac{d}{dt} \sin y\right)^2 = \dot{y}^2$$

so that $\frac{d}{dt} (\ddot{y} + \sin y) = 0$ can be expressed as $\dot{y}^2 \ddot{y}^2 + (y^{(3)})^2 = \dot{y}^2$. We have proved that y satisfies a differential algebraic equation. This elimination can be done in every differential equation where the coefficient satisfy a differential algebraic equation.

Definition 10.14 An input-output system defined as above is said to be a system of distributed parameters if and only if the differential fields are not ordinariesonce,doce.

Definition 10.15 If u is a transcendence base of $K\langle u, y\rangle / K$ then the input is said to be independent [4–6].

In other words, if $u \subset K\langle u\rangle \subset K\langle u, y\rangle$, then

$$\text{diff tr d}^\circ K\langle u, y\rangle / K = \text{diff tr d}^\circ K\langle u, y\rangle / K\langle u\rangle + \text{diff tr d}^\circ K\langle u\rangle / K$$

therefore diff tr d$^\circ$ $K\langle u, y\rangle / K\langle u\rangle = 0$.

10.4 Identifying the Input and Output of Dynamic Systems

For simple dynamic systems is easy to identify the input and the output, although this situation does not always occur, for example in complex electrical circuits, is necessary to decide which voltage and current variables are the inputs and outputs of the system.

Definition 10.16 Let K a differential field. A systemic field is a differential extension K/k (finitely generated) if there exists a finite set $w = (w_1, \ldots, w_s)$ of differential quantities such that $K = k\langle w\rangle$.

Example 10.7 Let $\dot{w} = w$ and $\mathbb{R}\langle w\rangle = \mathbb{R}\langle e^t\rangle$ the set of rational functions of e^t.

Definition 10.17 Let $m = \text{diff tr d}^\circ K/k$, where $0 \le m \le s$. There exists a non-unique partition of w in $w = (u, y)$ where $u = (u_1, \ldots, u_m)$ is a differential transcendence basis of K/k. $y = (y_1, \ldots, y_p)$, such that $m + p = s$ is formed of the remaining components. Since $K\langle u, y\rangle / K\langle u\rangle$ is differentially algebraic, it is said that u is the input and y the output.

Example 10.8 Let the integrator system $\dot{y} = u$. The inverse system (a perfect derivative) is also an input-output system.

10.5 Invertible Systems

One of the classical problems in automatic control is trajectory tracking of an output, there are two problems to solve. The first one is to know if the trajectory is reproducible. If so, the other problem is to calculate the inputs. One solution is through constant linear transfer matrices, in such way the problem is equivalent to left (or right) invertibility of matrices.

For linear systems with state-space representation, there is a fundamental algorithm called structure algorithm (of Silverman) that provides the response elements. In the nonlinear case, the answer is easy with SISO systems. For MIMO systems there is not an clear answer for the problem.

Definition 10.18 The differential rank of output ρ of system $K\langle u, y\rangle/\langle u\rangle$ is defined [10–12] by

$$\rho = \text{diff tr d}^\circ K\langle y\rangle/k$$

Two important properties of differential rank are listed below.

1.
$$\rho \leq \inf(m, p)$$

where m is the number of inputs and p is the number of outputs. It is clear that

a. $\rho \leq p$
b. $\rho \leq m$

since $k \subset k\langle y\rangle \subset k\langle u, y\rangle$. Furthermore,

$$\text{diff tr d}^\circ k\langle u, y\rangle/k = \text{diff tr d}^\circ k\langle u, y\rangle/k\langle y\rangle + \text{diff tr d}^\circ k\langle y\rangle/k$$

2. In a linear system, the differential rank of output is defined by a transfer matrix and is equal to the rank of this matrix.

Remark 10.2 The differential rank of output ρ is also the maximum number of outputs that are not related by any polynomial differential equation with coefficients in a field K (independent of x and u).

In practice, in certain simples cases, it is determined by looking for all the relationships of the form:

$$\bar{h}(y_1, \ldots, y_p) = 0, \qquad \bar{h} \in K\langle y_1, \ldots, y_p\rangle$$

considering a nonlinear system represented by

$$\Sigma : \begin{cases} \dot{x} = f(x) + g(x)u \\ y = h(x) \end{cases} \tag{10.1}$$

with $u = (u_1, \ldots, u_m) \in \mathbb{R}^m$, $x = (x_1, \ldots, x_n) \in \mathbb{R}^n$, $y = (y_1, \ldots, y_p) \in \mathbb{R}^p$

If we find r polynomials that are independent, then the differential output rank of the system (10.1) is

$$\rho = \text{diff tr d}^\circ K\langle y\rangle/K = p - r \tag{10.2}$$

Example 10.9

1. Let us define Σ_1, with two inputs and two outputs:

$$\Sigma_1 : \begin{cases} \dot{x}_1 = u_1 \\ \dot{x}_2 = u_1 \\ \dot{x}_3 = u_2 \\ y_1 = x_1 \\ y_2 = x_2 \end{cases}$$

it is possible to immediately observe that the only relationship of the type $p(y) = 0$ is a differential polynomial with coefficients in K given by:

$$p(y) = \dot{y}_1 - \dot{y}_2 = 0$$

a single-output is differentially independent and the differential rank of output ρ is

$$\rho = p - r = 2 - 1 = 1$$

2. Let us define Σ_2, with two inputs and two outputs:

$$\Sigma_2 : \begin{cases} \dot{x}_1 = u_1 \\ \dot{x}_2 = u_2 \\ \dot{x}_3 = u_2 \\ y_1 = x_1 \\ y_2 = x_2 + x_3 \end{cases}$$

there is no polynomial differential equation that relates the two outputs that are differentially algebraically independent so that the differential rank of output of the system is

$$\rho = p = 2$$

For a nonlinear system it is possible to associate two differential sets, namely polynomial ring and an ideal, i.e., given the following nonlinear system

$$\text{NLS} : \begin{cases} \dot{x} = f(x) + g(x)u \\ y = h(x) \end{cases} \tag{10.3}$$

we have that the polynomial ring and the ideal associated with the previous system is:

$$K \{u, x, f(x), g(x), h(x), y\} = K \{\Sigma\}$$
$$I = \{\dot{x} - f(x) - g(x)u, y - h(x)\} \subset K \{\Sigma\}$$

If we form the quotient ring with $K\{\Sigma\}$ and I, then we have the quotient ring $K\{\Sigma\}/I$ where the ideal I can be prime or not. If I is prime, we form the class quotient field of $K\{\Sigma\}/I$, i.e., $\mathcal{O}(K\{\Sigma\}/I)$. Then we extract a subfield $K\langle y\rangle$, with $K\langle y\rangle$ a differential extension of K, achieving this form the so called **differential rank of output** ρ given by

$$\rho = \text{diff tr d}^\circ\, K\langle y\rangle / K \tag{10.4}$$

If I is not prime the system is not well defined and f is ill-defined. We can show an example where ideal is not prime.

Example 10.10 Let us define the system $\bar{\Sigma}$ given by a nonlinear "implicit" system:

$$\bar{\Sigma} : \begin{cases} \dot{x}^2 = u^2 \\ y = x \end{cases}$$

We have $\bar{f}(x, \dot{x}, u) = \dot{x}^2 - u^2$ and $\bar{g}(x, y) = y - x$, then the associated differential sets are:

$$K\{x, u, y\} = K\{\bar{\Sigma}\}$$
$$I = \{\bar{f}, \bar{g}\} \subset K\{\bar{\Sigma}\}$$

Hence, $\bar{f}$ is factorizable, $\bar{f} = \dot{x}^2 - u^2 = (\dot{x} + u)(\dot{x} - u)$, that is to say $\bar{f}$ is not irreducible. Besides, $\dot{x}^2 - u^2 \in I$ but $\dot{x} - u \notin I$ and $\dot{x} + u \notin I$, thus the ideal I is not prime.

Definition 10.19 Let $K\langle u, y\rangle/\langle u\rangle$ a system with independent inputs.

1. The system is left invertible if and only if $\rho = m$.
2. The system is right invertible if and only if $\rho = p$.

Theorem 10.3 *The input-output system* $K\langle u, y\rangle/k\langle u\rangle$ *is left invertible if and only if*

$$\text{diff tr d}^\circ\, K\langle u, y\rangle / k\langle y\rangle = 0$$

$\square$

In other words, an input-output system is left invertible if and only if we can recover the input from the output by means of a finite number of differential equations.

Example 10.11 System $y = \dot{u}$ is left invertible because

$$u(t) = \int_0^t y(\sigma)\, d\sigma + c$$

On the other hand, $y = u_1 + u_2$ is not left invertible.

Definition 10.20 If the system $K\langle u, y\rangle/\langle u\rangle$ is square ($m = p$), left and right invertibility are equivalents. It is said that system is invertible.

10.6   Realization

One of the fundamental problems of automatic control is the realization or state space representation. Let L/K a finitely generated differential extension, i.e., there exists $w = (w_1, \ldots, w_s)$ such that $L = K\langle w \rangle$. The following conditions are equivalents:

1. The extension L/K is differentially algebraic.
2. Transcendence degree (non differential) of L/K is finite.

In other words, the non differential transcendence degree is the number of initial conditions necessary to calculate the solutions of differential algebraic equations.

Example 10.12

$$(\ddot{y})^2 - (\dot{y})^3 + y\dot{y} = 0$$

In this case, nondiff tr d° $= 2$.

Definition 10.21 Let u be a differential scalar indeterminate and let K be a differential field, with derivation denoted by $\frac{d}{dt}$. A dynamics is a finitely generated differential algebraic extension $G/K\langle u \rangle$, where $K\langle u \rangle$ denotes the differential field by K and the elements of a set $u = (u_1, u_2, \ldots, u_m)$ of differential quantities.

Remark 10.3 In the above definition, $G = K\langle u, \xi \rangle$ where $\xi \in G$. Any element of G satisfies a differential algebraic equation with coefficients that are rational functions over K in the components of u and a finite number of their time derivatives.

Example 10.13 The input-output system $\ddot{y} - uy = 0$ that is equivalent to the bilinear system:

$$\Sigma = \begin{cases} \dot{x}_1 = x_2 \\ \dot{x}_2 = ux_1 \\ y = x_1 \end{cases} \tag{10.5}$$

can be seen like a dynamics of the form $\mathbb{R}\langle u, y \rangle/\mathbb{R}\langle u \rangle$, where $G = \mathbb{R}\langle u, y \rangle$, $y \in G$, $K = \mathbb{R}$.

Definition 10.22 The input u is independent if and only if u is a differential transcendence base of K/k.

Since $K/k\langle u \rangle$ is a differential algebraic extension, the transcendence degree (non differential) of $K/k\langle u \rangle$ is finite, for example n. Let's take a finite set $\xi = (\xi_1, \ldots, \xi_\gamma)$, with $\gamma \geq n$ of elements in K, that contains a transcendence basis of $K/k\langle u \rangle$. The derivatives $\dot{\xi}_1, \ldots, \dot{\xi}_\gamma$ are $K\langle u \rangle$-algebraically dependent over ξ:

$$A_1\left(\dot{\xi}_1, \ldots, \dot{\xi}_\gamma, u, \dot{u}, \ldots, u^{(\alpha)}\right) = 0$$

$$\vdots$$

$$A_\gamma\left(\dot{\xi}_1, \ldots, \dot{\xi}_\gamma, u, \dot{u}, \ldots, u^{(\alpha)}\right) = 0$$

where $A_1, \ldots, A_\gamma$ are polynomials with coefficients in $k\langle u \rangle$. These equations are implicit due to implicit function theorem we obtain locally explicit differential equations:

$$\dot{\xi}_1 = a_1\left(\xi, u, \dot{u}, \ldots, u^{(\alpha)}\right)$$

$$\vdots$$

$$\dot{\xi}_\gamma = a_\gamma\left(\xi, u, \dot{u}, \ldots, u^{(\alpha)}\right)$$

Definition 10.23 ξ is said to be a generalized state of the dynamics $K/k\langle u \rangle$ and its dimension γ. A nominal state is a state of minimal dimension such that is a transcendence base of $K/k\langle u \rangle$ with dimension n. This state is characterized by the $k\langle u \rangle$-algebraic independence of its components.

10.7 Generalized Controller and Observer Canonical Forms

Let L/K be a finitely generated differential algebraic extension. The differential primitive element theorem states the existence of one element $\delta \in L$ such that $L = K\langle \delta \rangle$. On the other hand, let $K/k\langle u \rangle$ a nonlinear dynamic. Let suppose that $k\langle u \rangle$ is not a field of constants, for example, when the set of controllers u is not empty. Let $\delta \in K$ a differential primitive element of $K/k\langle u \rangle$. Consider the sequence of derivatives of δ: $\left(\delta, \dot{\delta}, \ldots, \delta^{(\gamma)}, \ldots\right)$.

Lemma 10.2 *The set* $\left(\delta, \dot{\delta}, \ldots, \delta^{(\gamma)}, \ldots\right)$ *is* $k\langle u \rangle$-*algebraically dependent (respectively independent) if and only if* $\gamma \leq n - 1$ *(respectively* $\gamma \geq n$*).* $n = $ diff tr d° $K/k\langle u \rangle$.

Proof By contradiction. We suppose that $\gamma \leq n - 1$ then the set $\left(\delta, \dot{\delta}, \ldots, \delta^{(\gamma),\ldots}\right)$ is $k\langle u \rangle$-algebraically dependent, this implies that diff tr d° $K/k\langle u \rangle \leq n - 1$. In the same way for $k\langle u \rangle$-algebraically independent and $\gamma \geq n$, diff tr d° $K/k\langle u \rangle \geq n$. In both cases there is a contradiction. $\square$

$\left(\delta, \dot{\delta}, \ldots, \delta^{(\gamma)}, \ldots\right)$ is a transcendence base of $K/k\langle u \rangle$, then we write

$$\mathscr{C}\left(\delta^{(n)}, \ldots, \dot{\delta}, \delta, u, \dot{u}, \ldots, u^{(n)}\right) = 0$$

where $\mathscr{C}$ is a polynomial with coefficients in K. According to the last, $x_1 = \delta$, $x_2 = \dot{\delta}, \ldots, x_n = \delta^{(n-1)}$ is a minimal state of the dynamics $K/k\langle u \rangle$. This leads to:

$$\dot{x}_1 = x_2$$
$$\dot{x}_2 = x_3$$
$$\vdots$$
$$\dot{x}_{n-1} = x_n$$
$$\mathscr{C}\left(\dot{x}_n, x_n, \ldots, x_1, u, \dot{u}, \ldots, u^{(\delta)}\right) = 0$$

that represents global generalized controller canonical form. In particular, the local representation is:

$$\dot{x}_1 = x_2$$
$$\dot{x}_2 = x_3$$
$$\vdots$$
$$\dot{x}_{n-1} = x_n$$
$$\dot{x}_n = \mathscr{C}\left(x_1, \ldots, x_n, u, \dot{u}, \ldots, u^{(\delta)}\right)$$
$$y = x_1 = \delta$$

10.8 Linearization by Dynamical Feedback of Nonlinear Dynamics

The problem of exact linearization by a static state feedback [13, 14] of a nonlinear dynamic has been proposed by Brockett, Jakubczyk and Respodek, also, the contributions by Hunt, Su and Meyer consist in Lie brackets and vector fields. Recently, Charlet, Levine and Marino gave sufficient conditions for linearization by a dynamic state feedback of nonlinear systems using geometric techniques.

Taking the local generalized controller canonical form and considering the equality between $\mathscr{C}$ and a homogeneous polynomial l with coefficients in K, in the variables $x = (x_1, \ldots, x_n)$, $v = (v_1, \ldots, v_m)$ that consist in a new input and a finite number of derivatives of v:

$$\mathscr{C}\left(x, u, \dot{u}, \ldots, u^{(\delta)}\right) = l(x, v, \dot{v}, \ldots, v^{(\beta)}) = \sum_{j=1}^{n} f_j x_j + \sum_{i=0}^{\beta} g_i v^{(i)}, \quad (f_j < 0, f_j, g_i \in K)$$

Theorem 10.4 *u contains a nonempty set of independent control variables then the dynamic $K / k\langle u \rangle$ can be locally linearized by a dynamic state feedback.* $\qquad\square$

Example 10.14 Let $K = \mathbb{R}$ and consider the dynamic with a single-input:

$$\dot{\eta}_1 = \eta_2 + u$$
$$\dot{\eta}_2 = \eta_2 \eta_3$$
$$\dot{\eta}_3 = u \eta_1$$

where $\mathbb{R}\langle u, \eta \rangle / \mathbb{R}\langle u \rangle$. It is not difficult to check that η_1 is a differential primitive element. Let $x_1 = \eta_1$, $x_2 = \dot{\eta}_1$, $x_3 = \ddot{\eta}_1$. It follows that:

$$\dot{x}_1 = x_2$$
$$\dot{x}_2 = x_3$$
$$\dot{x}_3 = u x_1 x_2 - u^2 x_1 + \frac{(x_3 - \dot{u}_1)^2}{x_2 - u} + \ddot{u} = f_1 x_1 + f_2 x_2 + f_3 x_3 + g_1 v + g_2 \dot{v} + g_3 \ddot{v} \quad (f_i < 0)$$

Making $v = \dot{v} = \ddot{v} = 0$, we have the following system

$$\dot{x}_1 = x_2$$
$$\dot{x}_2 = x_3$$
$$\dot{x}_3 = u x_1 x_2 - u^2 x_1 + \frac{(x_3 - \dot{u}_1)^2}{x_2 - u} + \ddot{u} = \sum_{i=1}^{3} f_i x_i, \ f_i < 0$$

With the following dynamical control law

$$\ddot{u} = \sum_{i=1}^{3} f_i x_i - u x_1 x_2 + u^2 x_1 - \frac{(x_3 - \dot{u}_1)^2}{x_2 - u}$$

redefining

$$u_1 = u$$
$$u_2 = \dot{u}$$

we have the following system

$$\dot{u}_1 = u_2$$
$$\dot{u}_2 = \sum_{i=1}^{3} f_i x_i - u_1 x_1 x_2 + u_1^2 x_1 - \frac{(x_3 - u_2)^2}{x_2 - u_1}$$

Finally, the system is extended with a stabilizing dynamical controller as above is mentioned, i.e.:

$$\dot{x}_1 = x_2$$

$$\dot{x}_2 = x_3$$

$$\dot{x}_3 = u_1 x_1 x_2 - u_1^2 x_1 + \frac{(x_3 - u_2)^2}{x_2 - u_1} + \dot{u}_2$$

$$\dot{u}_1 = u_2$$

$$\dot{u}_2 = \sum_{i=1}^{3} f_i x_i - u_1 x_1 x_2 + u_1^2 x_1 - \frac{(x_3 - u_2)^2}{x_2 - u_1}$$

To exemplify the performance of the stabilizing dynamical controller, a simulation is conducted, for this purpose the previous system has initial conditions $x_1(0) = 3$, $x_2(0) = 1$, $x_3(0) = -3$, $u_1(0) = 0$ and $u_2(0) = 0$ (note that this set of values make the system unstable), the expression $\sum_{i=1}^{3} f_i x_i = f_1 x_1 + f_2 x_2 + f_3 x_3$ determine the placement of the system's poles which have to be chosen so the system is asymptotically stable. A simple method to obtain the values of f_i is to solve f_i in the equation $(s + p_1)(s + p_2)(s + p_3) = s^3 - f_3 s^2 - f_2 s - f_1$ where p_i is the desired location of the poles that make the system stable, the solution of the equality gives the next relation of f_i and p_i:

$$f_1 = -p_1 p_2 p_3$$

$$f_2 = -(p_1 p_2 + p_1 p_3 + p_2 p_3)$$

$$f_3 = -(p_1 + p_2 + p_3)$$

For this example the values $f_1 = -24$, $f_2 = -26$ and $f_3 = -9$ lead to the characteristic polynomial $(s + 2)(s + 3)(s + 4)$ that causes a stable behavior shown in the following image (Fig. 10.1).

The original system's states are reproduced with the inverse transformation:

$$\eta_1 = x_1$$

$$\eta_2 = x_2 - u$$

$$\eta_3 = \frac{x_3 - \dot{u}}{x_2 - u}$$

Causing the system to behave as follows (Fig. 10.2).

The states of the system converge to zero within a few seconds, now, to see the action of the dynamical controller it is turned off, making the system have the next dynamic (Fig. 10.3).

Clearly the system without the aid of the dynamical controller is unstable proving that the control law can successfully make the system asymptotically stable.

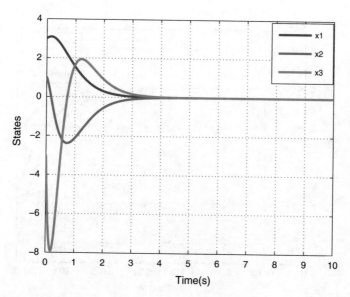

Fig. 10.1 Stabilized transformed system

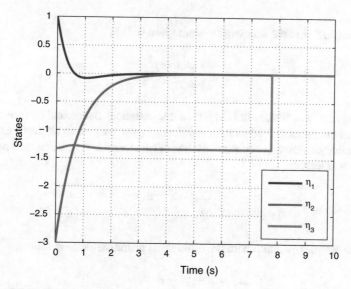

Fig. 10.2 Stabilized original system

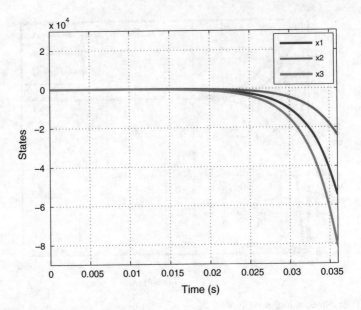

Fig. 10.3 Unstable transformed system

Example 10.15 Let the following system given in [15]

$$\dot{x} = x + y^3$$
$$\dot{y} = u$$

The system was treated in [15] with a discontinuous dynamical control law that was based on a smooth feedback for stabilization, yet, we give an alternative solution for the stabilization using a continuous dynamical control law. A differential primitive element is selected:

$$x_1 = x - u$$
$$\dot{x}_1 = x_2$$

A few simple operation with the differential primitive element follow:

$$x_2 = x + y^3 - \dot{u}$$
$$= x_1 + u + y^3 - \dot{u}$$
$$y = (x_2 - x_1 - u + \dot{u})^{1/3}$$

On the other hand

$$\dot{x} = \dot{x}_1 + \dot{u}$$
$$= x_2 + \dot{u}$$

Then

$$\dot{x}_2 = \ddot{x} + 3y^2\dot{y} - \ddot{u}$$
$$= (x_2 + \dot{u}) + 3u(x_2 - x_1 - u + \dot{u})^{2/3} - \ddot{u}$$

Making the transformed system be

$$\dot{x}_1 = x_2$$
$$\dot{x}_2 = (x_2 + \dot{u}) + 3u(x_2 - x_1 - u + \dot{u})^{2/3} - \ddot{u}$$

The intention is to make the system have a linear dynamic given by:

$$\begin{bmatrix} \dot{x}_1 \\ \dot{x}_2 \end{bmatrix} = \begin{bmatrix} 0 & 1 \\ a & b \end{bmatrix} \begin{bmatrix} x_1 \\ x_2 \end{bmatrix} \quad , a, b < 0$$

Choosing as dynamic controller

$$\ddot{u} = (x_2 + \dot{u}) + 3u(x_2 - x_1 - u + \dot{u})^{2/3} - ax_1 - bx_2$$

The system has the desired linear dynamic, the poles are obtained with the method presented in the previous example, a simulation is conducted using the parameters: $a = -3$, $b = -5$, $x_1(0) = 5$, $x_2(0) = -4$ So the system has the stable dynamic found in the image below (Fig. 10.4).

The original systems states are obtained by the inverse transformation:

$$x = x_1 - u$$
$$y = (x_2 - x_1 - u + \dot{u})^{1/3}$$

Which also have a stable dynamic (Fig. 10.5).

Another alternative to achieve the linearization is to select the transformation as:

$$x_1 = x$$
$$\dot{x}_1 = x_2$$
$$\dot{x}_2 = \ddot{x}$$
$$\dot{x}_2 = \dot{x} + 3y^2\dot{y}$$
$$= x + y^3 + 3y^2\dot{y}$$

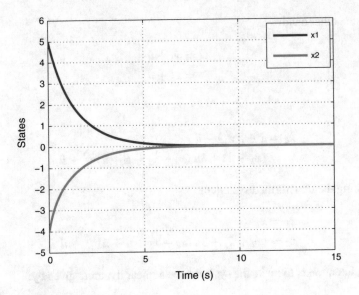

Fig. 10.4 Stabilized transformed system

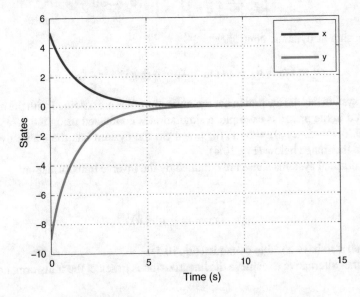

Fig. 10.5 Stabilized original system

Notice that this transformation does not yield a dynamic control law, but it does produce a smooth feedback. Since $y = (\dot{x} - x)^{1/3} = (x_2 - x_1)^{1/3}$

$$\dot{x}_2 = x_1 + (x_2 - x_1)^{3/3} + 3u\,(\dot{x} - x)^{1/3}$$

The transformed system is:

$$\dot{x}_1 = x_2$$
$$\dot{x}_2 = x_2 + 3u\,(x_2 - x_1)^{1/3}$$

Making the smooth control law be:

$$u = \frac{ax_1 + bx_1 - x_2}{3(x_2 - x_1)^{1/3}}$$

The system has the same linear dynamic, and employing the same parameters $a = -3$, $b = -5$, $x_1(0) = 5$, $x_2(0) = -4$ the next results are achieved (Fig. 10.6). Again the original systems states are obtained by the inverse transformation:

$$x = x_1$$
$$y = (x_2 - x_1)^{1/3}$$

That also produces a stable dynamic (Fig. 10.7).

This alternative choice of primitive element does not yield a dynamic feedback control law requiring various integrations, and yet it produces results on pair with the dynamic control, still it presents a singularity, so the implementation must be carefully made, specially when selecting initial conditions. To avoid infinite or undefined values that makes the original system or the transformed system unstable.

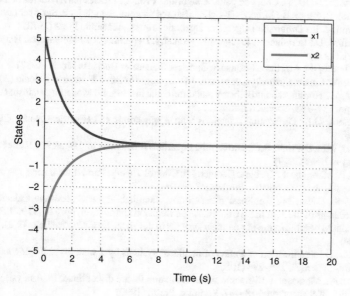

Fig. 10.6 Stabilized transformed system

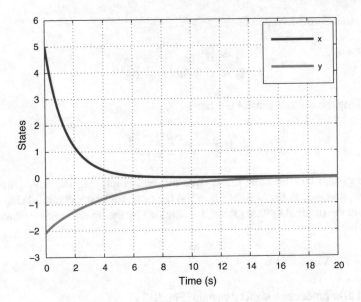

Fig. 10.7 Stabilized original system

References

1. Byrnes, C.I., Falb, P.L.: Applications of algebraic geometry in system theory. Am. J. Math. **101**(2), 337–363 (1979)
2. Dieudonn, J. (1974). Cours de gomtrie algbrique (Vol. 2). Presses universitaires de France
3. Samuel, P., Serge, B.: Mthodes d'algbre abstraite en gomtrie algbrique. Springer (1967)
4. Kolchin, E. R.: Differential algebra & algebraic groups. Academic Press (1973)
5. Mumford, D.: Introduction to algebraic geometry. Department of Mathematics, Harvard University (1976)
6. Hartshorne, R.: Algebraic Geometry. Springer Science & Business Media (1977)
7. Borel, A.: Linear algebraic groups, vol. 126. Springer Science & Business Media (2012)
8. Fliess, M.: Nonlinear control theory and differential algebra. In: Modelling and Adaptive Control, pp. 134–145. Springer, Berlin (1988)
9. Simmons, G.F.: Differential Equations with Applications and Historical Notes. CRC Press (2016)
10. Boyce, W.E., DiPrima, R.C.: Differential Equations Elementary and Boundary Value Problems. Willey & Sons (1977)
11. Falb, P.: Methods of Algebraic Geometry in Control Theory: Part I. Scalar Linear Systems and Affine Algebraic Geometry, Birkhuser (1990)
12. Diop, S., Fliess, M.: Nonlinear observability, identifiability, and persistent trajectories. In: Proceedings of the 30th IEEE Conference on Decision and Control, pp. 714–719. IEEE (1991)
13. Kawski, M.: Stabilization of nonlinear systems in the plane. Syst. Control Lett. **12**(2), 169–175 (1989)
14. Aeyels, D.: Stabilization of a class of nonlinear systems by a smooth feedback control. Syst. Control Lett. **5**(5), 289–294 (1985)
15. Fliess, M., Messager, F.: Vers une stabilisation non linaire discontinue. In: Analysis and Optimization of System, pp. 778–787. Springer, Berlin (1990)

Appendix
Important Tools

In this appendix, we collect some tools that are necessary in the development of some algebraic properties.

A.1 Summation

The addition of elements $a_1 + a_2 + a_3 + a_4 + a_5 + \cdots + a_n$ can be represented in a short form using the symbol Σ, that is:

$$\sum_{i=m}^{n} a_i = a_m + a_{m+1} + a_{m+2} + \cdots + a_{n-1} + a_n$$

where i is the variable of summation (or index), m is the lower bound of summation, n the upper bound of summation and $i = m$ means that the index i starts out equal to m. This index is incremented by 1 for each successive term and stopping when $i = n$.

The summation is the addition of a sequence of numbers that can be integers, rational, real or complex numbers. Note that the sum does not depend of this index and can be replaced for another index $\left(\sum\limits_{i=1}^{n} a_i = \sum\limits_{l=1}^{n} a_l \right)$. As examples of sums we have:

$$\sum_{j=1}^{10} x_j = x_1 + x_2 + x_3 + x_4 + x_5 + x_6 + x_7 + x_8 + x_9 + x_{10}$$

$$\sum_{j=1}^{4} 2^j \cdot j^2 = 2^1 \cdot 1^2 + 2^2 \cdot 2^2 + 2^3 \cdot 3^2 + 2^4 \cdot 4^2 = 2 + 16 + 72 + 256 = 346$$

© Springer Nature Switzerland AG 2019
R. Martínez-Guerra et al., *Algebraic and Differential Methods for Nonlinear Control Theory*, Mathematical and Analytical Techniques with Applications to Engineering, https://doi.org/10.1007/978-3-030-12025-2

Summation $\sum_{i=1}^{n} a_i$ can be established by mathematical induction, considering that

$$\sum_{k=1}^{0} x_k = 0 \tag{A.1}$$

$$\sum_{i=1}^{n+1} a_i = \sum_{i=1}^{n} a_i + a_{n+1} \tag{A.2}$$

Some important properties of summation are the following:

1. **Partition of summation.** If $m \in \{1, \ldots, n\}$ then

$$\sum_{i=1}^{n} a_i = \sum_{i=1}^{m} a_i + \sum_{i=m+1}^{n} a_i \tag{A.3}$$

2. **Additive property of summation.**

$$\sum_{i=1}^{n} (a_i + b_i) = \sum_{i=1}^{n} a_i + \sum_{i=1}^{n} b_i \tag{A.4}$$

3. **Homogeneous property.**

$$\sum_{i=1}^{n} (\lambda a_i) = \lambda \sum_{i=1}^{n} a_i \quad \forall \lambda \in \mathbb{R} \tag{A.5}$$

4. **Exchange of finite summations.**

$$\sum_{i=1}^{m} \sum_{j=1}^{n} a_{i,j} = \sum_{j=1}^{m} \sum_{i=1}^{n} a_{i,j} \tag{A.6}$$

5. **Finite summations of constants.**

$$\sum_{i=1}^{n} \lambda = \lambda n \quad \forall \lambda \in \mathbb{R} \tag{A.7}$$

6. **Linearity of the summation.**

$$\sum_{i=1}^{n} (\lambda a_i + \bar{\lambda} b_i) = \lambda \sum_{i=1}^{n} a_i + \bar{\lambda} \sum_{i=1}^{n} b_i \quad \forall \lambda, \bar{\lambda} \in \mathbb{R} \tag{A.8}$$

7. Telescoping summation.

$$\sum_{i=1}^{n} (a_{i+1} - a_i) = a_{n+1} - a_1, \quad \sum_{i=1}^{n} (a_i - a_{i+1}) = a_1 - a_{n+1} \tag{A.9}$$

Example A.1 The use of the properties mentioned above is shown in this exercise. Let consider:

$$a_n = \begin{cases} \frac{4^{2n+1} \cdot 2^{3n-2}}{8^{n-1}} & \text{if } 1 \leq n \leq 27 \\ \sqrt{2n+2} - \sqrt{2n} & \text{if } 28 \leq n \leq 56 \\ 7\left(n + n^2\right) & \text{if } 57 \leq n \end{cases}$$

and calculate the following sum:

$$\sum_{k=1}^{59} a_k$$

According to (A.3):

$$\sum_{k=1}^{59} a_k = \sum_{k=1}^{27} a_k + \sum_{k=28}^{56} a_k + \sum_{k=57}^{59} a_k$$

For the first addend, making algebraic reductions, taking account the homogeneous property (A.5) and the sum of the first n terms of a geometric series we have:

$$\sum_{k=1}^{27} a_k = \sum_{k=1}^{27} \frac{4^{2k+1} \cdot 2^{3k-2}}{8^{k-1}} = \sum_{k=1}^{27} \frac{4^{2k} \cdot 4 \cdot 2^{3k} \cdot 2^{-2}}{8^k \cdot 8^{-1}} = \frac{4 \cdot 2^{-2}}{8^{-1}} \sum_{k=1}^{27} \frac{4^{2k} \cdot 2^{3k}}{8^k} = 8 \sum_{k=1}^{27} 16^k$$

$$= 8 \cdot \frac{16\left(1 - 16^{27}\right)}{1 - 16} = -\frac{128\left(1 - 16^{27}\right)}{15}$$

On the other hand, for the second addend using the telescoping summation (A.9):

$$\sum_{k=28}^{56} a_k = \sum_{k=28}^{56} \left(\sqrt{2k+2} - \sqrt{2k}\right) = \sum_{k=28}^{56} \left(\sqrt{2(k+1)} - \sqrt{2k}\right) = \sqrt{114} - 2\sqrt{14}$$

and finally for the third addend:

$$\sum_{k=57}^{59} a_k = \sum_{k=57}^{59} 7\left(k + k^2\right) = 7 \sum_{k=57}^{59} k(k+1) = 7\left[57(58) + 58(59) + 59(60)\right] = 71876$$

Therefore,

$$\sum_{k=1}^{59} a_k = -\frac{128\left(1 - 16^{27}\right)}{15} + \sqrt{114} - 2\sqrt{14} + 71876$$

Sometimes it is necessary the change of variable in a summation, for example, let consider the following summation:

$$\sum_{k=2}^{5} a_k = a_2 + a_3 + a_4 + a_5$$

In the above expression, let $l = k + 2$ then $k = l - 2$. If $k = 2$ and $k = 5$ then $l = 4$ and $l = 7$ respectively, then we have

$$\sum_{l=4}^{7} a_{l-2} = a_2 + a_3 + a_4 + a_5 = \sum_{k=2}^{5} a_k$$

In addition as example, note that

$$\sum_{j=3}^{6} p^{j+2} = p^5 + p^6 + p^7 + p^8 = \sum_{m=5}^{8} p^m$$

The first summation, can be expressed as the second summation taking account the change of variable $m = j + 2$, then $j = m - 2$ and the lower and upper bound of the summation are $m = 5$ and $m = 8$ respectively. To conclude this section, see what happens with the cancellation of addends as special case of partition of a summation (A.3). Let the summation

$$\sum_{j=m}^{m+n} a_j + \sum_{j=l}^{l+k} a_j = a_m + a_{m+1} + a_{m+2} + \cdots + a_{m+n} - \left(a_l + a_{l+1} + a_{l+2} + \cdots + a_{l+k}\right)$$

where $l > m$, $l + k > m + n$. Let define the sets of index $\hat{A} := \{m, m + 1, m + 2, \ldots, m + n\}$, $\hat{B} = \{l, l + 1, l + 2, \ldots, l + k\}$. Suppose that from some i and even a $\hat{i}$:

$$a_{m+i} + a_{m+i+1} + a_{m+i+2} + \cdots + a_{m+\hat{i}} = a_{l+i} + a_{l+i+1} + a_{l+i+2} + \cdots + a_{l+\hat{i}}$$

then, the summation is reduced and these addends are eliminated (because each one is the additive inverse of the other). To determine which addends are in the final result, note that $\hat{C} := \hat{A} \cap \hat{B}$ represents a new set whose elements are the index of the addends to be canceled in the final result, that is,

$$\hat{A} \cap \hat{B} = \left\{m + i, m + i + 1, m + i + 2, \ldots, m + \hat{i}\right\} = \left\{l + i, l + i + 1, l + i + 2, \ldots, l + \hat{i}\right\}$$

Finally the index of addends of the final result are $\hat{A} \setminus \hat{C}$ for the summation $\displaystyle\sum_{j=m}^{m+n} a_j$

and $\hat{B} \setminus \hat{C}$ for the summation $\displaystyle\sum_{j=l}^{l+k} a_j$.

Example A.2 Consider the following summation

$$\mathscr{S} = \sum_{j=0}^{7} a_j - \sum_{j=1}^{11} a_j$$

According to theory discussed above

$$\hat{A} = \{0, 1, 2, 3, 4, 5, 6, 7\}$$

$$\hat{B} = \{1, 2, 3, 4, 5, 6, 7, 8, 9, 10, 11\}$$

then the set of indices of the addends to be canceled in the final result is $\hat{C} = \hat{A} \cap \hat{B} = \{1, 2, 3, 4, 5, 6, 7\}$ and $\hat{A} \setminus \hat{C} = \{0\}$, $\hat{B} \setminus \hat{C} = \{8, 9, 10, 11\}$ represent the sets of indices of the addends in the final result for the first and second addends respectively. Finally:

$$\mathscr{S} = \sum_{j=0}^{7} a_j - \sum_{j=1}^{11} a_j = a_0 - a_8 - a_9 - a_{10} - a_{11} = a_0 - \sum_{j=8}^{11} a_j.$$

A.2 Kronecker Delta

The Kronecker delta is a function of two variables, usually just positive integers. This function $\delta : \mathbb{Z} \times \mathbb{Z} \to \{0, 1\}$ is a piecewise function of variables i and j given by:

$$\delta_{i,j} = \begin{cases} 1, & \text{if } i = j, \\ 0, & \text{otherwise} \end{cases} \tag{A.10}$$

for example $\delta_{1,1} = 1$, $\delta_{10,5} = 0$, $\delta_{1,2} = 0$. In addition, for every $i, j \in \mathbb{Z}$, $\delta_{i,j} = \delta_{j,i}$. Some properties of Kronecker delta are listed below. These properties are useful in the simplification of summations with Kronecker delta.

1. $\displaystyle\sum_{i=1}^{n} \delta_{i,j} a_i = a_j$ for $j \in \{1, \dots, n\}$.

2. $\displaystyle\sum_{i=1}^{n} \delta_{i,j} a_i = 0$ for $j \notin \{1, \dots, n\}$.

3. $\displaystyle\sum_{k=1}^{n} \delta_{i,k}\delta_{k,j} = \delta_{i,j}.$

Example A.3 Considering the following summations

$$\sum_{i=1}^{5} \delta_{i,2}a_i = \delta_{1,2}a_1 + \delta_{2,2}a_2 + \delta_{3,2}a_3 + \delta_{4,2}a_4 + \delta_{5,2}a_5$$

$$= \overset{0}{\delta_{1,2}a_1} + \overset{1}{\delta_{2,2}a_2} + \overset{0}{\delta_{3,2}a_3} + \overset{0}{\delta_{4,2}a_4} + \overset{0}{\delta_{5,2}a_5} = a_2$$

In this case, according to property 1, $j = 2$ and $j = 2 \in \{1, 2, 3, 4, 5\}$, then we can use this property and avoid to make all the summation. In the following summation, $j = 6$. This value not belongs to set of indices, therefore the property 2 is used.

$$\sum_{i=1}^{5} \delta_{i,6}a_i = \delta_{1,6}a_1 + \delta_{2,6}a_2 + \delta_{3,6}a_3 + \delta_{4,6}a_4 + \delta_{5,6}a_5$$

$$= \overset{0}{\delta_{1,6}a_1} + \overset{0}{\delta_{2,6}a_2} + \overset{0}{\delta_{3,6}a_3} + \overset{0}{\delta_{4,6}a_4} + \overset{0}{\delta_{5,6}a_5} = 0$$

Finally,

$$\sum_{k=1}^{5} \delta_{i,k}\delta_{k,j} = \delta_{i,1}\delta_{1,j} + \delta_{i,2}\delta_{2,j} + \delta_{i,3}\delta_{3,j} + \delta_{i,4}\delta_{4,j} + \delta_{i,5}\delta_{5,j}$$

This summation is 1 if $i = j$ and belongs to set of indices or 0 if $i \neq j$ or $i = j$ but not belongs to the set of indices. In other words, this summation is the Kronecker delta $\delta_{i,j}$.

Kronecker delta is commonly used in matrix theory to define the identity matrix. This section conclude with other example where the entries of two matrices are written considering an explicit formula taking account the Kronecker delta.

Example A.4

$$A = \left[6\delta_{i,3}\right]_{i,j=1}^{4} = \begin{bmatrix} 6\delta_{1,3} & 6\delta_{1,3} & 6\delta_{1,3} & 6\delta_{1,3} \\ 6\delta_{2,3} & 6\delta_{2,3} & 6\delta_{2,3} & 6\delta_{2,3} \\ 6\delta_{3,3} & 6\delta_{3,3} & 6\delta_{3,3} & 6\delta_{3,3} \end{bmatrix} = \begin{bmatrix} 0 & 0 & 0 & 0 \\ 0 & 0 & 0 & 0 \\ 6 & 6 & 6 & 6 \end{bmatrix}$$

On the other hand, if $B \in \mathcal{M}_5(\mathbb{R})$ and $B_{i,j} = \delta_{i,4} + \delta_{j,2}$ then

$$B = \begin{bmatrix} \delta_{1,4}+\delta_{1,2} & \delta_{1,4}+\delta_{2,2} & \delta_{1,4}+\delta_{3,2} & \delta_{1,4}+\delta_{4,2} & \delta_{1,4}+\delta_{5,2} \\ \delta_{2,4}+\delta_{1,2} & \delta_{2,4}+\delta_{2,2} & \delta_{2,4}+\delta_{3,2} & \delta_{2,4}+\delta_{4,2} & \delta_{2,4}+\delta_{5,2} \\ \delta_{3,4}+\delta_{1,2} & \delta_{3,4}+\delta_{2,2} & \delta_{3,4}+\delta_{3,2} & \delta_{3,4}+\delta_{4,2} & \delta_{3,4}+\delta_{5,2} \\ \delta_{4,4}+\delta_{1,2} & \delta_{4,4}+\delta_{2,2} & \delta_{4,4}+\delta_{3,2} & \delta_{4,4}+\delta_{4,2} & \delta_{4,4}+\delta_{5,2} \\ \delta_{5,4}+\delta_{1,2} & \delta_{5,4}+\delta_{2,2} & \delta_{5,4}+\delta_{3,2} & \delta_{5,4}+\delta_{4,2} & \delta_{5,4}+\delta_{5,2} \end{bmatrix} = \begin{bmatrix} 0 & 1 & 0 & 0 & 0 \\ 0 & 1 & 0 & 0 & 0 \\ 0 & 1 & 0 & 0 & 0 \\ 1 & 2 & 1 & 1 & 1 \\ 0 & 1 & 0 & 0 & 0 \end{bmatrix}$$

A.3   Matrices

In this part we consider important a resume of properties of matrices and determinants. Consider the matrices $A, B, C \in \mathcal{M}_{m \times n}(\mathbb{F})$ and $\alpha, \beta \in \mathbb{F}$. In addition $0 = 0_{m \times n}$ represents the zero matrix.

1. **Sum of matrices and multiplication of one scalar by a matrix.**

   a. $A + B = B + A$
   b. $A + B + (C) = (A + B) + C$
   c. $\alpha(A + B) = \alpha A + \alpha B$
   d. $(\alpha + \beta)A = \alpha A + \beta A$
   e. $\alpha(\beta A) = (\alpha \beta)A$
   f. $A + 0 = A$
   g. $A + (-A) = 0$

2. **Multiplication of matrices.**

   a. $A(B + C) = AB + AC$
   b. $(A + B)C = AC + BC$
   c. $A(BC) = (AB)C$
   d. $\alpha(AB) = (\alpha A)B = A(\alpha B)$
   e. $A0 = 0A = 0$
   f. $BI = IB = B$
   g In general, $AB \neq AB$.
   h $AB = 0$ does not necessarily mean that $A = 0$ or $B = 0$.
   i $AB = AC$ does not necessarily mean that $B = C$.

3. **Trace properties.**
   The trace of a square matrix $A \in \mathcal{M}_{n \times n}(\mathbb{F})$ is defined to be the sum of the elements on the main diagonal of A:

   $$\text{tr}(A) = a_{11} + a_{22} + a_{33} + \cdots + a_{nn} = \sum_{i=1}^{n} a_{ii} \qquad (A.11)$$

   a. $\text{tr}(A + B) = \text{tr}(A) + \text{tr}(B)$.
   b. $\text{tr}(\alpha A) = \alpha \, \text{tr}(A)$.
   c. $\text{tr}(AB) = \text{tr}(BA)$.
   d. $\text{tr}(A^\mathsf{T}) = \text{tr}(A)$.
   e. $\text{tr}(A^\mathsf{T}B) = \text{tr}(AB^\mathsf{T}) = \text{tr}(B^\mathsf{T}A) = \text{tr}(BA^\mathsf{T}) = \sum_{i,j} A_{ij} B_{ij}$.
   f. $\text{tr}(ABCD) = \text{tr}(BCDA) = \text{tr}(CDAB) = \text{tr}(DABC)$. Trace is invariant under cyclic permutations.
   g. In general $\text{tr}(ABC) \neq \text{tr}(ACB)$.
   h. If A, B, C are symmetric matrices then $\text{tr}(ABC) = \text{tr}(A^\mathsf{T}B^\mathsf{T}C^\mathsf{T}) = \text{tr}(A^\mathsf{T}(CB)^\mathsf{T}) = \text{tr}((CB)^\mathsf{T}A^\mathsf{T}) = \text{tr}((ACB)^\mathsf{T}) = \text{tr}(ACB)$.

i. The trace is similarity-invariant: $\operatorname{tr}(P^{-1}AP) = \operatorname{tr}(P^{-1}(AP)) = \operatorname{tr}((AP)$ $P^{-1})) = \operatorname{tr}(A(PP^{-1})) = \operatorname{tr}(AI) = \operatorname{tr}(A)$.

j If A is a symmetric matrix and B is antisymmetric, then $\operatorname{tr}(AB) = 0$.

k If A is an idempotent matrix then $\operatorname{tr}(A) = \operatorname{rank}(A)$.

l. If A is a nilpotent matrix then $\operatorname{tr}(A) = 0$.

m. $\operatorname{tr}(I_n) = n$.

n. If $f(x) = (x - \lambda_1)^{d_1}(x - \lambda_2)^{d_2} \cdots (x - \lambda_k)^{d_k}$ is the characteristic polynomial of A, then $\operatorname{tr}(A) = d_1\lambda_1 + d_2\lambda_2 + \cdots + d_k\lambda_k$.

o. If $A \in \mathcal{M}_{n \times n}(\mathbb{F})$ and $\lambda_1, \ldots, \lambda_n$ are the eigenvalues of A, $\operatorname{tr}(A) = \sum_{i=1}^{n} \lambda_i$.

p. $\operatorname{tr}(A^k) = \sum_{i=1}^{n} \lambda_i^k$, where λ_i is an eigenvalue of A.

q. Let A^* the conjugate transpose of A, then $\operatorname{tr}(AA^*) \geq 0$. In addition $\operatorname{tr}(AA^*) = 0$ if and only if $A = 0$.

r. $\operatorname{tr}(B^*A) = \langle A, B \rangle$ defines an inner product on the space of matrices $\mathcal{M}_{m \times n}(\mathbb{F})$.

s. $0 \leq \operatorname{tr}(AB)^2 \leq \operatorname{tr}(A^2)\operatorname{tr}(B^2) \leq \operatorname{tr}(A)^2\operatorname{tr}(B)^2$.

4. **Diagonal matrices properties**.

If A, B are diagonal matrices then

a. $A + B = \operatorname{diag}(a_{11} + b_{11}, a_{22} + b_{22}, \ldots, a_{nn} + b_{nn})$.

b. $AB = \operatorname{diag}(a_{11}b_{11}, a_{22}b_{22}, \ldots, a_{nn}b_{nn})$.

c. $\alpha A = \operatorname{diag}(\alpha a_{11}, \alpha a_{22}, \ldots, \alpha a_{nn})$.

d. A is invertible if and only if the entries $a_1, \ldots, a_n$ are all non-zero, then $A^{-1} = \operatorname{diag}(a_1^{-1}, \ldots, a_n^{-1})$.

e. The adjoint of a diagonal matrix is again diagonal.

f. A square matrix is diagonal if and only if it is triangular and normal.

g. $\det(A) = \prod_{i=1}^{n} a_{ii}$.

5. **Properties of the inverse matrix**.

a. A^{-1} is unique.

b. $(A^{-1})^{-1} = A$.

c. $(AB)^{-1} = B^{-1}A^{-1}$

d. $(\alpha A)^{-1} = \alpha^{-1}A^{-1} \quad \forall \alpha \neq 0$.

e. $(A^n)^{-1} = (A^{-1})^n$.

f. $(A^{\mathsf{T}})^{-1} = (A^{-1})^{\mathsf{T}}$.

g. $A^{-1} = \frac{1}{\det(A)} \operatorname{adj} A$.

h. $\det(A^{-1}) = (\det(A))^{-1}$.

i. A matrix that is its own inverse, i.e. $A = A^1$ and $A^2 = I$, is called an **involution**.

6. **Transpose matrix properties**.
   a. $(A^\mathsf{T})^\mathsf{T} = A$.
   b. $(A + B)^\mathsf{T} = A^\mathsf{T} + B^\mathsf{T}$.
   c. $(AB)^\mathsf{T} = B^\mathsf{T} A^\mathsf{T}$.
   d. $(\alpha A)^\mathsf{T} = \alpha A^\mathsf{T}$.
   e. $\det(A^\mathsf{T}) = \det(A)$.
   f. Let $A \in \mathcal{M}_{n \times n}(\mathbb{F})$ then $\sigma(A) = \sigma(A^\mathsf{T})$, where σ denotes the spectrum of a matrix.
   g. Let $A \in \mathcal{M}_{n \times n}(\mathbb{R})$. A is called a **symmetric matrix** if $A^\mathsf{T} = A$.
   h. Let $A \in \mathcal{M}_{n \times n}(\mathbb{R})$. A is called a **skew-symmetric matrix** if $A^\mathsf{T} = -A$.
   i. Let $A \in \mathcal{M}_{n \times n}(\mathbb{R})$. A is called a **orthogonal matrix** if $A^\mathsf{T} = A^{-1}$. In addition, $AA^\mathsf{T} = A^\mathsf{T}A = I$.
   j. Let $A \in \mathcal{M}_{n \times n}(\mathbb{C})$. A is called a **Hermitian matrix** if $A^\mathsf{T} = A^*$.
   k. Let $A \in \mathcal{M}_{n \times n}(\mathbb{C})$. A is called a **skew-Hermitian matrix** if $A^\mathsf{T} = -A^*$.

7. **Symmetric and skew-symmetric matrices properties**.
   Let $A \in \mathcal{M}_{n \times n}(\mathbb{F})$.

   a. $A + A^\mathsf{T}$ is a symmetric matrix.
   b. AA^T is a symmetric matrix.
   c. $A - A^\mathsf{T}$ is a skew-symmetric matrix.

   If A, B are symmetric (skew-symmetric) matrices:

   a. $A + B$ is a symmetric (skew-symmetric) matrix.
   b. αA is a symmetric (skew-symmetric) matrix.
   c. AB is not necessarily a symmetric (skew-symmetric) matrix.

8. **Conjugate transpose properties**.
   Let $A^* = \left(\bar{A}\right)^\mathsf{T} = \bar{A}^\mathsf{T}$ the conjugate transpose or Hermitian transpose, where A^T denotes the transpose matrix and $\bar{A}$ is matrix with complex conjugated entries. Then

Index

© Springer Nature Switzerland AG 2019
R. Martínez-Guerra et al., *Algebraic and Differential Methods for Nonlinear
Control Theory*, Mathematical and Analytical Techniques with
Applications to Engineering, https://doi.org/10.1007/978-3-030-12025-2